The iPhone PocketGuide

Christopher **Breen**

All the Secrets of the iPhone, Pocket Sized.

Peachpit Press

The iPhone Pocket Guide
Christopher Breen

Peachpit Press
1249 Eighth Street
Berkeley, CA 94710
510/524-2178
510/524-2221 (fax)

Find us on the Web at: www.peachpit.com
To report errors, please send a note to errata@peachpit.com

Peachpit Press is a division of Pearson Education

Editor: Jim Akin
Copy editor: Kathy Simpson
Production editor: Tracey Croom
Compositor: David Van Ness
Indexer: Rebecca Plunkett
Cover design: Aren Howell
Cover photography: Scott Cowlin
Interior design: Kim Scott, with Maureen Forys

ISBN 13: 978-0-321-51008-2
ISBN 10: 0-321-51008-9

9 8 7 6 5 4 3 2 1

Printed and bound in the United States of America

Dedication

As always, to my little iBreen, Addie.

About the author

Christopher Breen has been writing about Apple's efforts and technology since the latter years of the Reagan Administration for such publications as *MacUser*, *MacWEEK*, and *Macworld*. He currently pens *Macworld's* Mac 911 tips and troubleshooting column and is the author of *The iPod and iTunes Pocket Guide*.

Acknowledgments

This book would be just another unrealized intriguing idea if not for the dedication of the following individuals.

At Peachpit Press: Cliff Colby who, mere minutes after the iPhone's announcement, sidled up and said "So, I suppose you'd like to write a book about this;" Jim Akin, who spent as many late nights editing the book as I did writing it; Kathy Simpson who acted as Copy Editor Plus™, providing both expert eyes to grammar, word-choice, and layout, but also the eagle-eyed "Uh, you said exactly the opposite thing in Chapter 2;" and Production Pro Tracey Croom who patiently put up with my "No, we are *not* going to take photographs of the iPhone. You're going to have to wait until I figure out how to hack the phone to pull screenshots from it."

At home: My wife, Claire, who not only put up with the late hours, missed weekends, and working-vacation, but picked up my iPhone, played with it for a couple of minutes, and said, "I want one of these. I can't wait for your book to come out." And my daughter Addie who gave her Dada a huge smile and welcome hug at the end of every working day.

Abroad: Ben Long for hacking the iPhone so we could include screenshots of the interface; Erica Sadun for creating the utility that made the screenshots possible; Jason Snell at *Macworld* who never said "I'd like exclusive rights to that brain full of iPhone goodness;" *Macworld's* Kelly Turner, Jon Seff, Dan Frakes, Rob Griffiths, Jim Dalrymple, and Dan Moren for sharing their iPhone experiences with me; and the boys from System 9 for their continued cool cattedness.

And, of course, the sleep-deprived designers, engineers, and other Apple folk who gave birth to the iPhone. Congratulations!

Contents

Getting Started

I really, *really* appreciate your purchasing this book. And because I do, it would be rude of me to delay, even for an instant, the pleasure you'll gain from using your cool new iPhone. Allow me to offer a few quick steps for firing up your new best friend:

1. Open the box.

   I know it's a cool box, but it stands between you and your prize. When you open the box, peel away the plastic that wraps your iPhone. Lift out the plastic tray, and remove the Dock connector cable and the Dock.

2. Download iTunes 7.3 or later.

   If you don't already have the latest version of iTunes, travel to www.apple.com/itunes and grab a copy. Versions are available for Windows PCs as well as the Mac.

3. String the cable from your computer to the Dock.

   Attach the USB end of the cable to a free USB 2.0 port on your Mac or PC. Plug the Dock connector into the back of the Dock. Then place your iPhone in the Dock's cradle.

tip **Use one of the ports on your computer or a powered USB 2.0 hub. A USB port on your keyboard won't provide enough power to charge the iPhone.**

4. Walk through the AT&T and iTunes setup process.

   With your computer on, the cable and Dock attached, and the iPhone in the Dock's cradle, iTunes should launch automatically. When it does, it walks you through the activation process. (The iPhone won't work until it's activated.)

   The activation steps you take depend on your current status as a mobile-phone user. You'll follow separate paths if you're already an AT&T Wireless subscriber, if you're transferring a number from another cellular carrier, or if you're a first-time wireless customer.

   As you move through the setup process, you'll be asked to choose a service plan. The chief differences among AT&T's plans are the amount of talk minutes you have each month and, of course, the price—three plans offered at $60, $80, and $100 per month for 450, 900, and 1,350 talk minutes, respectively. All AT&T iPhone plans include unlimited data and 200 Short Message Service (SMS) text messages per month.

Other than the iPhone, computer, and copy of iTunes, the other things you'll need are a Social Security number or an AT&T preapproved credit-check code (issued by an AT&T representative) and a credit card number or existing Apple ID. (Your credit card won't be charged anything.)

5. Leave your phone on, and wait for it to activate.

Presuming that you got the A-OK at the end of the activation initiated by iTunes, AT&T should send a signal to your iPhone to activate it. Before the phone is activated, it displays a message saying that it's waiting for activation.

note **While you wait, you're welcome to play with all of the iPhone features that aren't focused on communications (and which won't work until the phone is activated). Specifically, you can't yet use the Web browser, email, or YouTube, but you can play with the camera, imported photos, and iPod features.**

6. Make a call.

I know you're eager to load some music and video on your iPhone, but first you should call someone and brag about having the coolest phone on the planet. If the device has been plugged in for an hour, you should have enough juice to brag with the best of 'em.

tip **If the battery is low, and you plan to brag a long time, unplug the Dock connector cable from the back of the Dock and jack it into the bottom of the iPhone; the phone will charge while you talk.**

Now tap the Phone icon in the bottom-left corner of the iPhone's screen, tap the Keypad icon at the bottom of the resulting screen, tap in a number on the keypad that appears, and tap Call. Brag.

7. Configure iTunes.

   Put the iPhone back in the Dock, and it appears in iTunes' Source list. Select it, and the iPhone preferences pane fills most of iTunes' main window.

   Note the Info, Music, Photos, Podcasts, and Video tabs. These tabs are where you choose what to sync to your iPhone. This procedure is just complicated enough that after you poke around in iTunes, I suggest that you move to the next step:

8. Read the rest of this book.

Everything you need to know to make the most of your iPhone is within its pages. Enjoy your stay!

1

Meet the iPhone

On January 9, 2007, Apple, Inc.'s Steve Jobs took the stage at San Francisco's Moscone West convention center to open the annual Macworld Expo. Close to the 30-minute mark of his presentation, he moved, with these words, to the main subject of the day:

"Today, we're introducing three revolutionary products of this class. The first one is a widescreen iPod with touch controls. The second is a revolutionary mobile phone. The third is a breakthrough Internet communications device.

"A widescreen iPod, a mobile phone, and a breakthrough Internet communications device," he repeated ... and repeated again until the crowd finally caught on with a roar.

Rather than describing three separate devices, Jobs was introducing one: the iPhone.

The iPhone is all that and more:

- Indeed, it is the first widescreen video-capable iPod, featuring a unique interface for browsing your music collection by album cover and a beautifully bright 3.5-inch display where you can view pictures, TV shows, music videos, and movies in a widescreen way.
- As a mobile phone, the iPhone offers many of the finest features of today's sophisticated phones (and a few features these "smart" phones haven't even thought of), offering such options as speakerphone, conference calling, SMS messaging, contact syncing with a computer or Internet service, and a visual voice-mail scheme that allows you to choose quickly just those messages you want to listen to.
- And as an Internet communications device, the iPhone offers a full-blown browser capable of displaying real Web pages, email, maps, helpful applications for quickly getting information on weather and stocks, and even streamed YouTube videos.

And more? Yes, lots more. To begin with, the beauty of the iPhone is that these three major elements work hand in hand. It's the work of a moment to search for a nearby pizza joint in the Maps application, click its contact information link, and call to order your dinner, for example. Better yet, these elements operate so intuitively that for perhaps the first time in your life, you'll use *all* the features your mobile phone offers, rather than just those that don't require you to spend hours with a poorly written manual. Add to that the iPhone's unique multitouch screen display, which lets you use natural finger motions and virtual controls that appear on the iPhone's screen to control your phone; a sensor that detects the phone's vertical or horizontal orientation and rotates its images accordingly; built-in Wi-Fi and Bluetooth capability; and a 2-megapixel digital camera—and you've got a fairly formidable hunk of technology in your pocket.

Oh, and did I mention that the iPhone works with both Windows PCs and Macs? Or that the computer application that handles the handshake between your computer and the iPhone is one you're already familiar with? Yes, that would be the same iTunes you now use to load your iPod with music, podcasts, and movies.

In this inaugural chapter, I look at the items that come in the iPhone box, as well as the physical features and controls that make up this three-in-one wonder.

Boxed In

The squat black box holds more than the iPhone. Within, you'll find these goodies.

iPhone

Well, of course. You didn't lay out $500 or $600 with the dream of getting an electric shaver, did you? Lift the top of the box, and the iPhone is the first thing you see, suspended in its clear plastic tray.

Stereo headset

Yes, you can press the iPhone against your face just as you can any other telephone, but if you do, you miss the opportunity to use your iPhone to check your stocks or surf the Web as your mother complains, for the 37th time, that the coot across the way has been nipping at her nasturtiums. The stereo headset, with its on-board microphone, not only frees your hands for playing with the iPhone's other features, but also lets you listen to your favorite tunes and serves as an audio aid when you're watching a movie.

The headset's mic—the small gray plastic cylinder found about 5 inches down the line from the right earbud—is also a switch. While listening to music or watching a video, press it once to pause playback. Press it twice to move to the next track when listening to music (pressing twice doesn't cause you to move to the next chapter in a movie). Press it once to answer a call and again to end the call. To decline a call, press and hold for a couple of seconds, and then let go. The iPhone will beep twice to acknowledge your action. While you're in the middle of a call, press once to answer an incoming call and put the first call on hold. To end the current call and answer an incoming call or switch to a call on hold, press and hold for 2 seconds, and then let go.

USB power adapter

If you purchased an iPod back in the day, this white plastic cube will look familiar to you. Smaller than an iPod charger, this gadget is the wall charger for your iPhone. One edge includes flip-down blades that you plug into a convenient socket. (The unit that holds these blades can be removed and replaced with an adapter for use in a different country.) The other edge has a female USB port.

Dock connector-to-USB cable

This cable is the one you string between the Dock connector port on the bottom of the iPhone or the Dock and either the USB power adapter or a USB port on your computer. When connected to a computer, this cable acts as both data and power link between the iPhone and computer. Without it, you can't sync media and information from the computer to the iPhone.

Dock

Although you could power and sync your iPhone with just the included USB cable, wouldn't you like a more attractive setting for your new jewel? The white plastic Dock is just that. When placed in the Dock, the iPhone leans back in a jaunty position.

The Dock's cradle includes a 30-pin male Dock connector that passes power and data through the female Dock connector port at the back of the Dock. The Line Out port on the Dock's back can connect your iPhone to amplified speakers or a stereo receiver. (When the iPhone is connected this way, its volume control has no effect on the volume. Rather, the device the Dock is plugged into—that stereo receiver, for example—controls the volume.)

note **No, those many holes in the Dock's cradle don't indicate that you received a defective unit. They're there to let sound out from the iPhone's small speaker as well as into the iPhone's microphone—both of which are on the bottom of the phone.**

Cleaning cloth

The iPhone's face is quite scratch resistant, but it's bound to pick up smudges. To keep it clean, use the included soft black cleaning cloth.

On the Face of It

Thanks to its touch-screen display, the iPhone sports very few buttons and switches. Those that it does possess, however, are important (**Figure 1.1**).

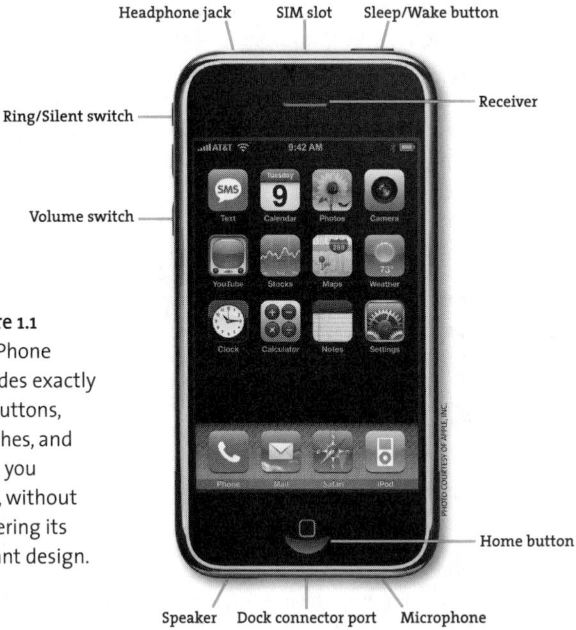

Figure 1.1 The iPhone provides exactly the buttons, switches, and ports you need, without cluttering its elegant design.

Up front

After peeling the plastic off your iPhone and flipping it in your hand a time or two, you will come to a remarkable realization: The thing apparently has but one button! No number keys, no tiny joystick, no Answer and Hang Up buttons—just an indented round button at the bottom of the display. This button is the Home button, and as its name implies, it takes you to the iPhone's Home screen nearly every time you press it. (OK, I'll end the suspense: You also use the Home button to wake up your sleeping phone. When you do, you don't go Home but, rather, after unlocking the phone, you see the last screen that was visible when the phone dozed off.)

The front of the iPhone also bears a small slit near the top of the device. This slit is the receiver—the hole from which you listen to the person you're speaking with when you operate the phone in traditional phone-to-face mode.

On top

Look a little more carefully, and you'll discover a few more mechanical controls and ports. On the top edge of the iPhone is a tiny black switch. Apple describes this switch as the Sleep/Wake button, which you also use to turn the iPhone on and off.

To lock the phone, press this button. (To unlock the phone, press the Home button and slide your finger where it says "Slide to Unlock.") To switch the phone off, press and hold the Sleep/Wake button for a few seconds until a red slider appears onscreen, labeled *Slide to Power Off.* Drag the slider to the right to switch off the phone (or tap Cancel to belay that order). "Drag?" you ask. Yes, the gesture is exactly what it sounds like. Place your finger on the Arrow button, and slide it to the right. (I describe all the iPhone's gestures in the "Full Gestures" section later in this chapter.)

To turn the phone on after shutting it off, press and hold the Sleep/Wake button until you see the Apple logo on the display.

Next to the Sleep/Wake button is a very thin strip of metal, which is the edge of the SIM (Subscriber Information Module) card, which holds a programmable circuit board that stores your personal subscriber information. This card allows the iPhone to work. Without it, you've got a pretty hunk of metal, glass, and acrylic. Unlike some other mobile phones, the iPhone comes with this removable card preinstalled.

Hunkered down in the top-left corner is the Headphone jack, which accommodates the iPhone's white headset plug. Although that headset bears the same 3.5mm connector used on modern headphones—the kind of headphones you use with your iPod—you're likely to discover that your headphones won't work with it. Apple has recessed this port for the sake of a clean design. The downside of this is that the plug on those expensive headphones of yours won't seat properly, thus ensuring silence when you plug them into the iPhone.

Down below

Smack-dab in the middle of the iPhone's bottom edge is the familiar-to-iPod-owners Dock connector port. This port is a proprietary 30-pin connector used for syncing the iPhone and attaching such accessories as power adapters, FM transmitters, and speaker systems.

The holes in the bottom-left corner of the iPhone are for the device's built-in speaker—used for both speakerphone and audio playback when nothing is plugged into the Headphone jack.

The holes in the bottom-right corner are for the phone's microphone.

To the left

Unlike some other mobile phones, the iPhone has a physical Ring/Silent switch for silencing the phone. You'll find this toggle switch on the top-left side of the phone. When the switch is toggled toward the face of the phone, the iPhone is in the Ring position. Push the switch toward the back of the phone to silence the ringer.

note

When the Ring/Silent switch is set in the Ring position, the iPhone will make noise when you receive a call, voice-mail message, text message, or email message; when an appointment or alarm-clock alarm goes off; when you lock the phone; and when you type with the iPhone's keyboard.

When the switch is set to Silent, the phone makes noise only when an alarm-clock alarm goes off.

Below the Ring/Silent switch is the Volume rocker switch. Press it up to increase volume on a call or when listening to music or watching a movie; press it down to decrease the volume.

The back

Other than the shiny Apple logo, the iPhone name, the iPhone's capacity, and some really tiny print, the only thing you'll find on the back of the phone is the camera lens.

note

And by "the only thing," I do mean that you won't find a lever, switch, or button to open the iPhone for the purpose of replacing its battery. Like the iPod, the iPhone does not offer a user-replaceable battery. When your iPhone's battery gives up the ghost, you must have it serviced. See Chapter 9 for more on the iPhone's battery.

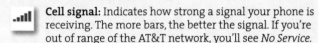

Icon See That

When operating your iPhone, you'll see a variety of small icons on its face. Here's what they mean:

Cell signal: Indicates how strong a signal your phone is receiving. The more bars, the better the signal. If you're out of range of the AT&T network, you'll see *No Service.*

Airplane mode: All functions that broadcast a signal—making a call, using Wi-Fi or EDGE networks, connecting to Bluetooth devices—are shut off when you switch the phone into airplane mode.

Wi-Fi: Indicates that you're connected to a Wi-Fi network. The stronger the signal, the more bars.

EDGE: Indicates that you're within range of AT&T's EDGE network.

Lock: Your phone is locked.

Play: Your iPhone is playing music.

Alarm: You've set an alarm.

Bluetooth: If you see a blue or white Bluetooth icon, the iPhone is linked to a Bluetooth device. If you see a gray Bluetooth icon, Bluetooth is on, but the phone's not linked to a Bluetooth device.

Battery: Indicates the battery level and whether the battery's charging. A white battery icon with a lightning bolt tells you that the phone is charging. When you see a green battery icon with a plug icon, the battery is fully charged.

Applications

Steve Jobs wasn't kidding when he claimed that the iPhone is a widescreen iPod, a mobile phone, and an Internet communications device. As I write these words, Apple bundles 15 applications with the iPhone. You access these applications from the iPhone's Home screen, which you can easily summon by pushing the Home button on the face of the phone.

The Big Four

The iPhone's four most powerful applications—the ones that act as the gateway to the device's phone, music, video, email, and Web browsing functions—appear at the bottom of the Home screen.

Phone

Tap the Phone icon on the Home screen, and you're taken to the main Phone screen, where you can make calls, pull up a list of your contacts, view recent calls, and listen to your voice mail. I describe this area in rich detail in Chapter 3.

Mail

This application is the iPhone's email client. As with the email client on your computer, you use the iPhone's Mail to compose and send messages, as well as read and manage received email. I look at Mail in Chapter 4.

Safari

Safari is Apple's Web browser. Unlike other mobile phones, the iPhone carries a real live Web browser rather than a "baby browser" that grudgingly allows you to view only a small portion of the material a Web page offers. When you pull up a Web page in the iPhone's Safari, it looks and behaves like a real Web page. Chapter 5 is devoted to Safari.

iPod

Perhaps the coolest iPod ever made is incorporated into your iPhone. Capable of playing both audio and video, the iPhone is a wonderful on-the-go media player. Look to Chapter 6 for the ins and outs of the iPod functions.

The Littler 11 (plus 1)

But the applications don't stop with the Big Four. The iPhone also includes smaller applications that handle things like text messaging, calendars, stocks, and weather.

Text

No, this app isn't a full-blown instant-messaging client, though it looks like one (specifically, like Mac OS X's iChat). This application is for sending and receiving SMS (Short Message Service) text messages. I look at the Text application in Chapter 3.

Calendar

When you sync your iPhone, you can transfer calendar events and alarms from Apple's iCal and from Microsoft's Entourage and Outlook. These transferred items appear in the iPhone's Calendar application. You can also add events directly to the phone by using the iPhone's keyboard, and then sync those events with your computer. I discuss Calendar in more detail in Chapter 4.

Photos

Tap the Photos icon, and you'll see a list of photo albums—the first holding the pictures you've taken with the iPhone's camera; the next, the complete collection of all synced photos on the iPhone; and then any albums or folders you've synced with the phone. Chapter 7 offers more details on the iPhone's photo capabilities.

Camera

Use this application to snap a picture with the iPhone's built-in 2-megapixel camera. From within the Camera application, you can email pictures you've taken, as well as use one of those pictures as a contact's icon. The camera and photos are the subjects of Chapter 7.

YouTube

With this application, you can view streamed YouTube videos on your iPhone. YouTube, being a visual-based application, is examined in Chapter 7.

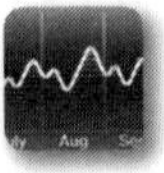

Stocks

Similar to the Stocks widget in Apple's Mac OS X, the iPhone's Stocks application lets you track your favorite stocks in near real time. All widgety things are detailed in Chapter 8.

Maps

Lost? A street map is just a tap away. Based on Google Maps, this application quickly provides not only maps, but also current driving conditions, satellite views, and the location of businesses within each map. Chapter 8 covers the Maps app.

Weather

Much like another Mac OS X widget, the Weather application displays current conditions, as well as the six-day forecasts for locations of your choosing. Like I said, Chapter 8 is great.

Clock

Find the time anywhere in the world, as well as create clocks of favorite locations. You also use the Clock application to create alarms and to invoke the stopwatch and countdown timer. Yeah, see Chapter 8 for this one too.

Calculator

Still can't figure out an appropriate tip without using your fingers? Pull up the iPhone's Calculator to perform common math operations. You're not going to make me write it again, are you? *Sigh*. OK, Chapter 8.

Notes

Notes is the iPhone's tiny text editor. Use the phone's virtual keyboard to create lists, jot down reminders, compose haiku, or remind yourself to look in Chapter 8 for more details.

Settings

Settings is the "plus one" application in this list. Though Settings technically isn't an application, a tap of the Settings icon produces a preferences window for configuring such features as airplane mode, Wi-Fi, usage, sounds, brightness, wallpaper, general settings (including date and time, auto-lock, password lock, network, Bluetooth, and keyboard), mail, phone, Safari, iPod, and photos. Though I'll discuss Settings in regard to specific applications throughout this little tome, I provide the big picture in Chapter 2.

Full Gestures

The iPhone's screen is deliberately touchy: Touching it is how you control the device. This section covers the gestures you use to navigate and control your phone.

Tap

You're going to see the word *tap* a lot in this book. When you want to initiate an action—launch an application, control the phone's iPod playback features, flip a object around, or move to the next screen—this gesture is the one you'll likely use.

Double-tap

Sometimes, just one tap won't do. Double-tapping often enlarges or contracts an image—zooms in on a photo or Web page, for example, or returns it to its normal size after you've enlarged it. Other times, it can make items return to the previous view. Double-tap a playlist in Cover Flow view, for example, to flip the playlist back to the album artwork.

Flick

If you want to scroll up or down a long list rapidly on your iPhone, zip through a selection of album covers in the iPhone's Cover Flow view (a view that allows you to browse your music and podcast collection by album cover/artwork), or flip from one photo to another, you use the flick gesture. As you flick faster, the iPhone attempts to match your action by scrolling or zipping more rapidly. Slower flicks produce less motion on the display.

To stop the motion initiated by a flick, just place your finger on the display. Motion stops instantly.

Drag

For finer control, drag your finger across the display. Use this motion to scroll in a controlled way down a list or email message, or to reposition an enlarged image or Web page. You also drag the iPhone's volume slider and playhead when you're in the iPod area.

Stretch/pinch

To expand an image—a photo or Web page—place your thumb and index finger together on the iPhone's display and then stretch them apart. To make an image smaller, start with your thumb and finger apart and then pinch them together.

Touch and drag

This gesture isn't a terribly common one but a helpful one nonetheless. In the iPod's More area, you'll find the option to swap out icons along the bottom of the display by touching and dragging new icons into place. You also touch and drag entries in the On-The-Go playlist to change their positions in the list.

Entering and Editing Text

Taps, pinches, and drags help you navigate the iPhone, but they won't compose email messages for you, help correct spelling mistakes, or delete ill-considered complaints. The iPhone's keyboard and a well-placed finger will do these jobs.

Touch typing

The iPhone's virtual keyboard largely matches the configuration of your computer's keyboard. You'll find an alphabetic layout when you open most applications (**Figure 1.2** on the next page). To capitalize characters, tap the upward-pointing arrow key (the iPhone's Shift key). To view numbers and most punctuation, tap the .?123 key. To see less-used characters (including £, ¥, and €), choose the numbers layout by tapping the .?123 key then tapping the #+= key. The Space, Return, and Delete keys do exactly what you'd expect.

Figure 1.2 The iPhone's keyboard.

To make typing easier, the keyboard's layout will change depending on the application you're using. In Mail, for example, the bottom row holds the @ symbol along with a period (.). While working in Safari the default layout will show period (.), slash (/), and .com keys along the bottom. And, unlike with any other iPhone application, you can display a keyboard in Landscape view in Safari, which gives you more room to type.

tip **When you type a character, its magnified image appears as you touch it. If you tap the wrong character, leave your finger where it is and slide it to the character you want—the character won't be "typed" until you let go of it.**

Editing text

The iPhone offers a unique way to edit text. You needn't tap the Delete key time and again to work your way back to your mistake. Instead, tap and hold on the line of text you want to edit. When you do, a magnifying glass appears (**Figure 1.3**), showing a close-up of the area under your finger. Inside this magnified view is a blinking cursor. Drag the cursor to where you want to make your correction—after the word or letter you want to correct—and then use the Delete key to remove the text. In most cases you can also tap between words to insert the cursor there.

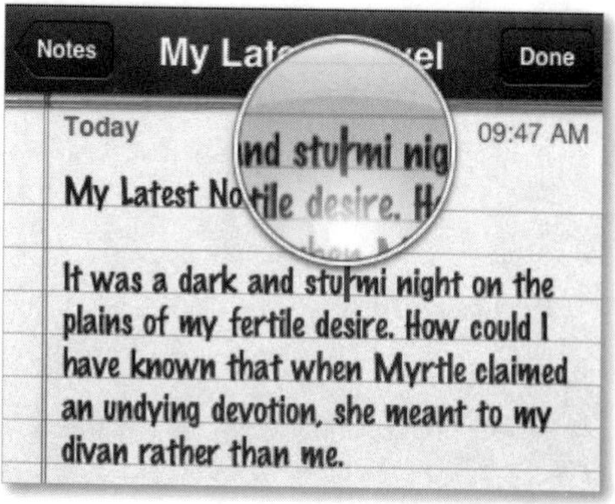

Figure 1.3 Tap and hold to magnify your mistakes.

2

Setup, Sync, and Settings

It would be great if your iPhone were ready to make calls, play music, and surf the Web right out of the box. Then again, it would be just as great if the groceries you purchased from the local Piggly Wiggly cooked themselves. But like those groceries, your iPhone needs a measure of preparation before it can do you any good.

This chapter is devoted to just that kind of preparation. It offers insight into how iTunes and the iPhone interact, and it covers the details of the iPhone's settings and preferences, starting at the point just after you've pulled the iPhone from the box.

Activation

If you've purchased a mobile phone in the past, you're likely accustomed to trotting down to a carrier's store; filling out reams of paperwork; and, in the process, promising the company your firstborn should you miss a monthly payment. Apple and AT&T have made activating your iPhone a far simpler process—one you can complete in the comfort of your own home or workplace.

Get iTunes

If you don't have a copy of iTunes 7.3 or later, now's the time to make a beeline for www.apple.com/itunes. The iPhone cannot be activated or synced without this version (or later) of iTunes, and Apple doesn't include it or any other software in the iPhone box.

iTunes is available in both Macintosh and Windows versions. For the iPhone to work with your Mac, you must be running Mac OS X 10.4.10 or later, and your Mac should have a 500 MHz G3 processor or better. PC users must be running Windows 2000, XP, or Vista on a 500 MHz Pentium processor or better.

Plug in the iPhone

Plug the included USB cable into a free USB 2.0 port on your Mac or PC. You can plug the data-connector end of the cable into the bottom of the iPhone, if you like, but if you plan to sync your iPhone to this computer routinely, why not use the included Dock as well? Plug the cable into the back of the Dock, and insert the iPhone into the Dock's cradle.

When you first plug in your iPhone, its screen displays the Apple logo. Then, after going dark for a short period, it displays an instruction to activate the phone.

At the bottom of the screen, you'll see the message *Slide for Emergency*. Don't worry—you haven't been seated in your iPhone's exit row. All mobile phones have an emergency call feature that allows you to dial 911 regardless of their carrier or activation status. If you experience severe chest pains due to the excitement of owning an iPhone, you may want to put this feature to good use; otherwise, leave it alone.

Use iTunes to activate AT&T account

iTunes acts as the handshake between your iPhone; your computer; and (during activation) the iPhone's mobile carrier, AT&T. That relationship begins at the very start—with the activation of your phone. Here's what to expect from iTunes.

iTunes should launch automatically. If you're launching this version of iTunes for the first time, you'll be asked to accept the license agreement (which, naturally, you won't read a word of).

When iTunes launches, the activation process begins—provided that you have an active Internet connection. Activation can't take place without such a connection.

The first thing you see is a reassuring screen that tells you what will happen: You'll activate your phone with AT&T; register and get an iTunes Store account (if you don't already have one); and put your contacts, music, and other content on your iPhone (**Figure 2.1**).

Figure 2.1
What's in store.

tip **In this screen, as in most screens throughout the registration process, you'll see a FAQ button. If at any time, you have questions about what iTunes is asking you to do, click this button for an explanation of what's going on.**

Who are you? The next screen ascertains whether you're an existing AT&T (formerly Cingular Wireless) customer or are new to AT&T. If you're an existing customer, you have the option of transferring a number in your existing account to the iPhone (which takes your old phone out of service), or adding a new line to your existing account and keeping your old phones and numbers in service.

If you choose the existing-customer option, you'll be asked to enter your current mobile-phone number, a billing zip code, and the last four digits of your Social Security number (AT&T has this number on record and uses it to verify your identity).

If you're a new customer, you can activate a single iPhone, or activate two or more iPhones (you lucky devil!) on an individual or FamilyTalk Plan. (In a FamilyTalk Plan, members of a family share all the talk minutes and SMS messages allotted to the account.) When you select the new-customer option, you can choose to transfer an existing mobile-phone number from your old phone and carrier to the iPhone. Just enter the old phone number; the account number from your old carrier (Verizon or T-Mobile, for example); your billing zip code; and, if applicable, your account password for your old carrier. If you don't have an existing mobile-phone number, click the Continue button without entering any information, and you'll be prompted to set up a new account.

When you choose to activate two or more phones, you can activate each on an individual account or activate all iPhones on a FamilyTalk Plan.

After you've made your choices and filled in any information asked of you, you'll be told to wait while AT&T verifies your account information. Under normal circumstances, this process takes about 1 minute.

Choose your plan. When your information is verified, you'll be asked to choose a data plan. If you're an existing AT&T mobile customer with a current calling plan, you can stay with that plan and simply agree to the iPhone's Data Plan, which runs $20 extra a month for unlimited data services (email and Web), Visual Voicemail (the iPhone system that lets you browse voice-mail messages in a list and listen to them in any order you like), and 200 SMS text messages each month (**Figure 2.2**). Unused SMS messages do not roll over to the next month, and each SMS message in excess of the 200-per-month allowance costs 5 cents.

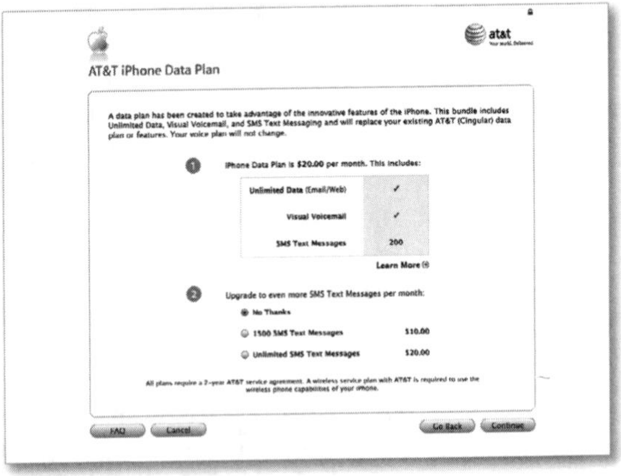

Figure 2.2 If you have an existing AT&T account, you simply add the Data Plan for $20 a month.

note **Each SMS message is limited to 160 characters. The iPhone plan includes an option to purchase 1,500 SMS messages for an extra $10 a month or unlimited SMS messages for $20 a month. But you can probably get a better deal: AT&T offers unlimited text messaging for $5 a month as I write this chapter. Call your AT&T representative to learn more about this option.**

If you don't have an existing plan, you're offered three plan choices. All of them include unlimited data, 200 SMS messages, and rollover minutes (a plan that stockpiles the minutes you don't use for a period of 12 months). The $60 plan includes 450 minutes of talk time and 5,000 night and weekend minutes. The $80 plan offers 900 talk minutes and unlimited nights and weekends. The $100 plan gives you 1,350 talk minutes and the same unlimited nights and weekends. If you like, you can purchase additional talk minutes.

note **When you sign up for a plan, you're committing to a 2-year contract. Cancellation fees will apply if you decide to leave AT&T before those 2 years are up.**

Get (or verify) an Apple ID. You're largely done with AT&T now. After you've provided your information to AT&T and chosen a plan, iTunes will ask you to verify your Apple ID or sign up for one if you don't already have one (**Figure 2.3**). If you have an Apple ID, simply enter it and its password when prompted.

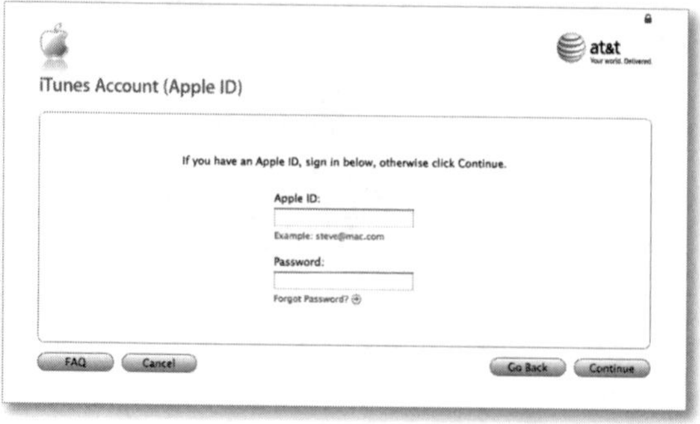

Figure 2.3 Create or verify an Apple ID.

If you don't have an Apple ID, click the Continue button without filling in the Apple ID and Password fields, and you'll be prompted to create an ID. To do so, you must type an email address. Figure 2.3 shows a mac.com address (steve@mac.com), but you can use any email address you have. You will also be asked to provide a credit card number. This request is a little confusing, as an Apple ID is free. But should you want to purchase something from the iTunes Store—music or movies, for example—Apple uses the credit card on record for the purchase. Regrettably, you cannot activate the phone without an Apple ID.

After you've entered or created an Apple ID, iTunes may take up to a minute for validation.

Accept, or else. The next couple of screens present the iPhone and AT&T service agreements. At the bottom of the screen, you'll see a check box indicating that you've read and accepted each agreement. Whether you've read and accepted or not, if you don't check this box, you can't activate the phone. You're welcome to print these agreements for when you find it hard to fall asleep; Apple's runs to just over eight pages and can be read comfortably with a magnifying glass.

Review your information, and activate. Before submitting your information for good, you have the chance to review it, including your contact information, the phone number your iPhone will have, and the details of your plan.

Click Submit, and AT&T will begin activating your phone. Under normal circumstances, activation can take up to 3 minutes (**Figure 2.4**).

Figure 2.4
Congrats! Your iPhone is nearly ready to use.

After AT&T has activated the phone at its end, it beams information to the iPhone to activate the device itself. Leave your phone on, and when it's ready to work, a message on its display will tell you so. In the meantime, you're welcome to set up the phone in iTunes.

Setting up the iPhone

After AT&T and Apple have finished with the phone, you can finally do something with it other than admire its shiny black face. What say we put something on your iPhone other than fingerprints?

Automatically sync personal information

Apple understands that despite all its cool media and Web features, the iPhone is a mobile phone. As such, it should do what other higher-end mobile phones do, such as store your contacts, calendars, email accounts, and Internet bookmarks. The first step of iTunes' iPhone setup—the one in which you have the opportunity to name your phone—makes this easy by offering to sync these items automatically (**Figure 2.5**).

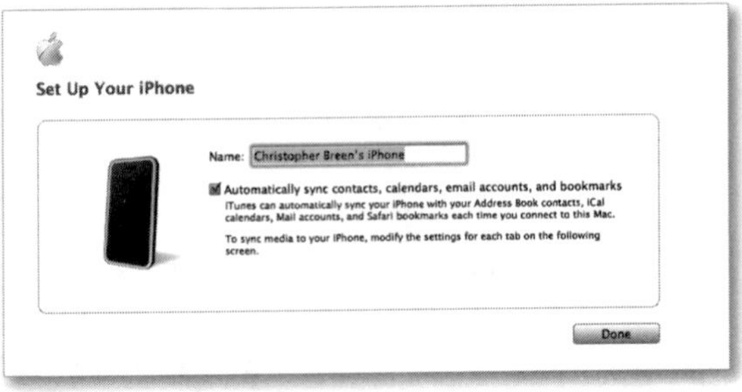

Set Up Your iPhone
Name: Christopher Breen's iPhone
Automatically sync contacts, calendars, email accounts, and bookmarks
iTunes can automatically sync your iPhone with your Address Book contacts, iCal calendars, Mail accounts, and Safari bookmarks each time you connect to this Mac.
To sync media to your iPhone, modify the settings for each tab on the following screen.
Done

Figure 2.5 Choose the Automatically Sync option to sync important data from your computer to your iPhone.

If you're using a Mac, iTunes will sync the iPhone with your Address Book contacts, iCal calendars, Apple Mail accounts, and Safari bookmarks. If you're using a Windows PC, it will sync contacts from Windows Address Book or Microsoft Outlook; calendars from Outlook; and email accounts from Windows Mail (included with Windows Vista), Outlook Express (Windows XP), or Outlook.

If you'd rather tell iTunes exactly what information to sync, you can do this later and in a more specific fashion. To choose the manual method, simply uncheck the Automatically Sync Contacts check box in this window and then click Done.

Tab-tastic

If you've synced a display-bearing iPod with your Mac or PC lately, you won't be startled by the iTunes iPhone Preferences window. Like the iPod Preferences window, this one contains a series of tabs for syncing data to the iPhone. Those tabs shake out as follows.

Summary

As its name suggests, the Summary tab provides an overview of your iPhone. Here, you'll find the iPhone's name (which you can change by clicking it in iTunes' Source list and entering a new name), its capacity, the software version it's running, its serial number, and its phone number (**Figure 2.6**).

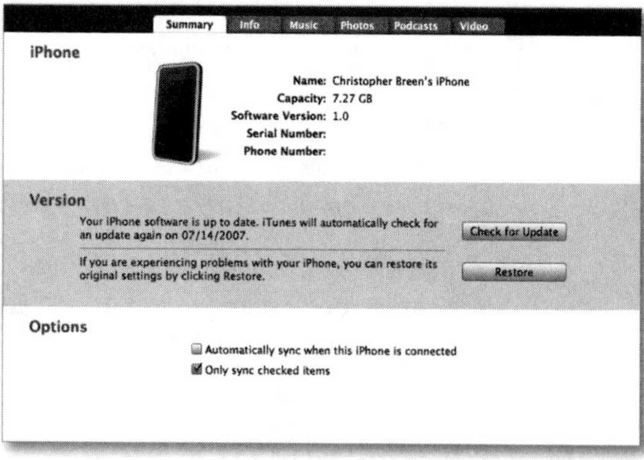

Figure 2.6 The iTunes Summary tab.

In the Version portion of the window, you learn whether your iPhone's software is up to date (you can make sure that you have the latest version by clicking the Check for Update button). Here, you also find a Restore button for placing a new version of the iPhone software on the device. (I revisit this button in Chapter 9.)

The Summary tab provides two selectable options. The first, Automatically Sync When This iPhone Is Connected, does exactly that: It tells iTunes to sync your phone whenever you dock it. If iTunes isn't running when you dock the iPhone, the phone launches automatically and starts syncing. Disable this option if you don't want iTunes to replace any of the iPhone's content automatically. This setting is carried with the phone, which means that regardless of which computer you jack it into, it does what this setting instructs.

tip **The iPhone panel within the iTunes Preferences window contains an option similar to this one: the Disable Automatic Syncing for All iPhones check box. The option in the Summary tab applies only to an individual iPhone; the iTunes setting applies to *all* iPhones. When you check Disable Automatic Syncing for All iPhones, no iPhone connected to the computer will sync.**

The other option—Only Sync Checked Items—was, frankly, confounding to me. Apple could have cleared up my confoundment by extending its label to read Only Sync Checked Items *in Your iTunes Library* because, you see, that's what it means. If you want greater control over what is planted on your iPhone, you can check some songs on a playlist in your iTunes Library, but not others. When this option is turned on, when you sync the iPhone, it will sync only the items you've checked.

Info

The Info tab is where you choose which data—contacts, calendars, mail accounts, and browser bookmarks—you'd like to sync to your iPhone (**Figure 2.7**). Contacts settings come first and, like iPod preferences, allow you to sync all your contacts or just selected groups you've created in Apple's Address Book (if you use a Mac), or in Windows Address Book or Outlook on a PC.

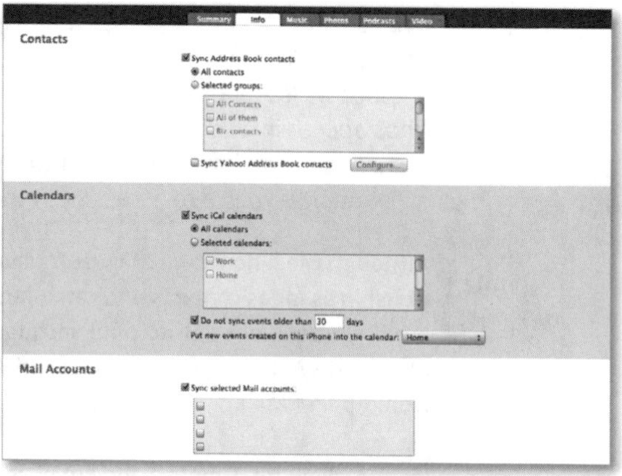

Figure 2.7 The iTunes Info tab.

You'll also see an option for syncing your online Yahoo Address Book contacts. Click the Configure button, agree to the license agreement, and enter your Yahoo username and password. When this option is enabled, your Yahoo Address Book contacts will make their way to the iPhone too.

The Calendars area works similarly. You can choose to sync all your calendars or, if you're using a Mac, just selected calendars you've created in Apple's iCal (your work calendar, for example, but not your personal calendar). Regrettably, the iPhone cannot maintain separate calendars. Events from all calendars you choose to sync appear within one calendar in the iPhone Calendar application, and there's no way to distinguish, say, work appointments from personal ones.

iTunes will look within Apple's Mail on a Mac and in Outlook, Microsoft Mail, and Outlook Express on a Windows PC for email account settings. Those it finds appear in a list in the Mail Accounts area of the Info tab. You have the option to select the email account(s) you'd like to access with the iPhone.

note **Although the iPhone and Microsoft have a cozy relationship as far as contacts and calendars are concerned, the iPhone will not pull account settings from Microsoft's Macintosh email client, Entourage—only from Apple's Mail.**

tip **If, after allowing iTunes to add email accounts to the Info tab automatically, you add a different account (maybe you've changed ISPs or taken on a .Mac account, for example), iTunes will add it to the Mail Accounts area automatically when you next sync your iPhone.**

The Mail/Entourage Relationship

Although the iPhone doesn't support pulling account information from Microsoft's popular Macintosh email client, Entourage 2004, it can sync Entourage contacts and calendars, but in an indirect way. Recent versions of Entourage include a new Sync Service feature, which you'll find in Entourage's Preferences window. Choose Sync Services, and you'll find the option to synchronize contacts with Apple's Address Book and .Mac. Likewise, you can synchronize Entourage events and tasks with iCal and .Mac. (Entourage's Notes, however, are not supported by the iPhone.) When this option is switched on, Address Book, iCal, and Entourage swap data as you add it.

Enable Sync Services, and any events and contacts you've stored in Entourage will be synced to your iPhone. Leave it off, and your Entourage data will remain missing in action.

In the Web Browser area of the Info window in the Macintosh version of iTunes, you'll find a Sync Safari Bookmarks check box. On a Windows PC, a pop-up menu provides the option to sync Safari or Internet Explorer bookmarks (**Figure 2.8**).

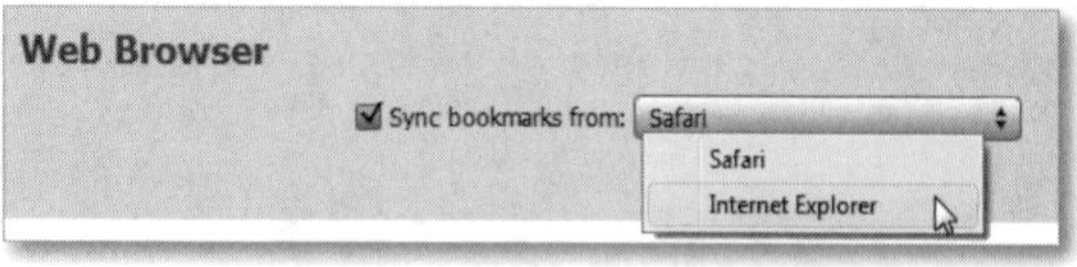

Figure 2.8 Windows users can sync both Safari and Internet Explorer bookmarks.

Music

The Music tab is exactly like the one in the iPod Preferences window (**Figure 2.9**). Enable the Sync Music option and then choose to sync All Songs and Playlists or just Selected Playlists (all the playlists in your iTunes Library appear in the list below). At the bottom of the window, you'll find the option to include music videos when you sync your iTunes music.

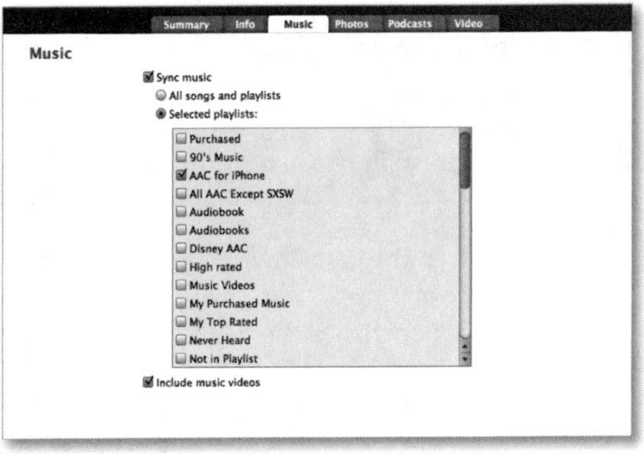

Figure 2.9
The Music tab.

In Chapter 6, I discuss how to sync music most efficiently.

Photos

If you use a Mac, the iPhone can sync photos with Apple's iPhoto and Aperture, as well as with your Photos folder or a different folder of your choosing (**Figure 2.10**). On a Windows PC, it can sync with your My Pictures folder, a different folder of your choosing, or photo albums created with Adobe Photoshop Elements 3.0 or later or Adobe Photoshop Album 2.0 or later.

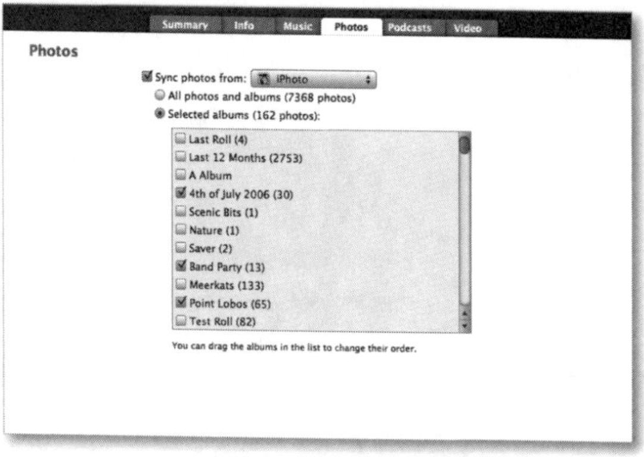

Figure 2.10 The iTunes Photos tab.

When syncing with an application that supports albums—iPhoto, for example—you can select specific albums you'd like to sync with. If you're syncing with a folder, you'll see the option to sync with specific folders within that folder.

I cover photos in rich detail in Chapter 7.

Podcasts

Just as with an iPod, you can listen to podcasts on your iPhone. Because people tend to listen to lots of podcasts, some of which tend to be long (their files therefore taking up significant amounts of room), iTunes lets you manage which ones are synced to your iPhone (**Figure 2.11** on the next page).

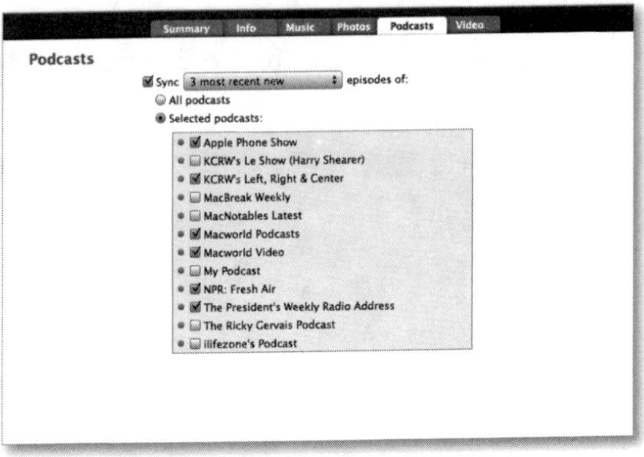

Figure 2.11 The iTunes Podcasts tab.

As in each one of these tabs, you have the option to not sync this content. But if you choose to, you can choose all or the 1, 3, 5, or 10 most recent podcasts you've downloaded from all the podcasts you have or just from selected podcasts. Or you can sync all or the 1, 3, 5, or 10 most recent unplayed podcasts—again, from all your podcasts or just selected podcasts. Finally, you can sync all or the 1, 3, 5, or 10 most recent unplayed podcasts—yes, from all your podcasts or just selected podcasts.

Note that video as well as audio podcasts are included here. Because video can consume a lot of storage space, be careful how you choose your video podcasts.

For more on podcasts, check Chapter 6.

Video

This tab is the last of the bunch and includes options for syncing TV Shows and Movies. The arrangement here is very much like the Podcasts tab. To help you sync exactly the video you want, both the TV Shows and Movies areas allow you to sync very specifically (**Figure 2.12**).

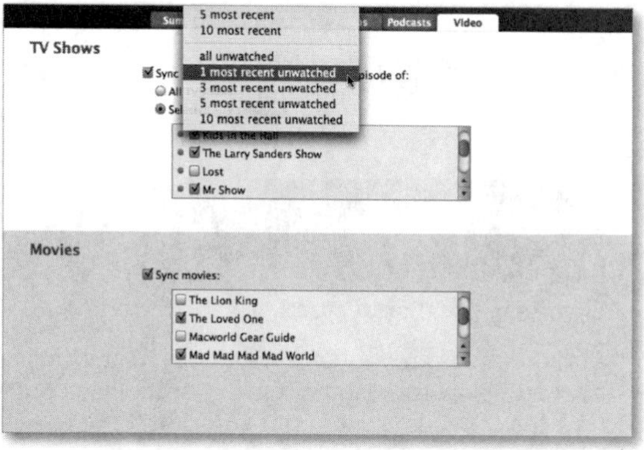

Figure 2.12 The iTunes Video tab.

In the TV Shows area, you'll see all the TV programs listed in the TV Shows category in iTunes' Source list. You can sync all or the 1, 3, 5, or 10 most recent TV shows among all TV shows or just those selected shows or playlists from the list of shows below. Or you can sync all unwatched or the 1, 3, 5, or 10 most recent unwatched episodes of all TV shows or selected TV shows or playlists.

The Movies area of the Video tab is very straightforward. You'll find a simple Sync Movies check box that you enable if you want to sync movies to the iPhone. Just select from the list the movies you want to sync. This list is derived from the movies that fall within the Movies category in iTunes' Source list.

Down Below

I would be remiss if I left the iPhone Preferences window without mentioning the Capacity bar at the bottom of the window (**Figure 2.13**). Familiar to iPod owners, this bar details how your iPhone's storage space is being used. Here, you view the total capacity of your iPhone, along with statistics for Audio, Video, Photos, Other (including contacts and calendars, for example), and Free Space.

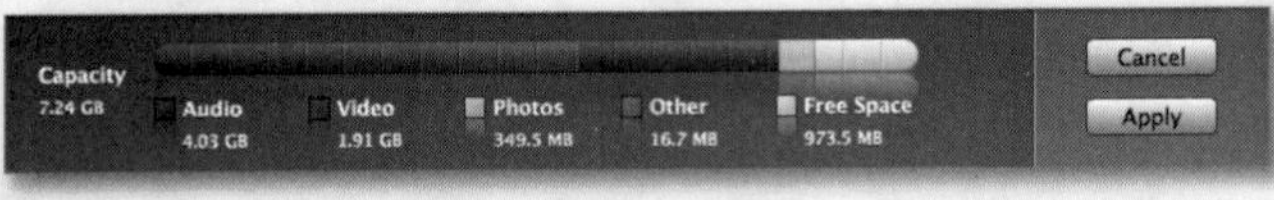

Figure 2.13 The iTunes Capacity bar.

By default, the amount of storage consumed by a particular item appears below its heading (Video 1.25 GB, for example). But if you click the Capacity bar, the statistics labels change—first to the number of items in each category and then, with another click, to the amount of time it would take to play all the videos and audio stored on the iPhone (2.5 days, for example).

To the right of the Capacity bar is the Sync button. Click this button to sync the iPhone right now rather than waiting for the next time you dock the thing.

If you make a change in your sync settings—change photo albums, for example, or choose a new movie or podcast to sync—the Sync button disappears, and Cancel and Apply buttons take its place. To sync the iPhone immediately with the new settings, click Apply. If you think better of your choices, click Cancel to undo your changes.

Settings

Although you'll control much of your iPhone's behavior within its applications, some global settings have some bearing on how it performs. You access these settings by tapping the Settings icon in the iPhone's Home screen. Here's what Settings contains (**Figure 2.14**).

Figure 2.14 The Settings screen.

Airplane Mode

Modern mobile phones are far more than just devices for annoying patrons in restaurants and movie theaters. Like the iPhone, they can play music and videos, as well as display pictures. Little good these features do you on a cross-country flight, however, when some of the phone's features interfere with an aircraft's navigation. Thus was born airplane mode, which switches off a phone's wireless features—calling, email, and web browsing, for example—but allows you to use the phone's other features.

This mode is a simple on/off setting. When it's on, you can't make or receive calls, use email, browse the Web, or use a Bluetooth accessory. You can continue to listen to music, watch videos, check your calendar, listen to your visual voice mail, view pictures, and read email and text messages stored on the phone.

tip

If you're interested in getting the longest media play time out of your iPhone, turn on airplane mode. Enabling the wireless features—even without using them—pulls power from your battery.

Wi-Fi

The iPhone supports 802.11 wireless networking. In this screen, you can turn Wi-Fi on or off. Turning it off saves some battery power. I describe the workings of the iPhone's Wi-Fi settings in the "Network" section later in this chapter.

Usage

If you're wondering how long it's been since your iPhone was fully charged or how many minutes you've yakked, here's where to look (**Figure 2.15**). The Usage screen includes information on when you last charged the iPhone fully, how long the iPhone has been in standby mode since the last full charge, the number of minutes you've talked in the current billing period and over the life of the phone, and how much data you've sent and received over the EDGE network.

Figure 2.15 The Usage screen.

To reset the current call time period and EDGE statistics, tap Reset Statistics. The time since last full charge and lifetime call-time statistics remain after a reset.

Sounds

In the Sounds screen (**Figure 2.16**), you choose among the iPhone's 25 ringtones. (Regrettably, as I write this book, you can't add your own ringtones.) You can adjust the volume of these rings by dragging the Volume slider just above the Ringtone entry.

Figure 2.16 The Sounds screen.

You also use the Sounds setting to determine which phone behaviors and events are assigned alert sounds and which aren't. You can set sounds to accompany the arrival of new voice mails, text messages, and emails; the successful sending of email; and calendar alerts, lock sounds, and keyboard clicks. By default, all these actions make sounds; you can turn these sounds off and on, but you can't customize them. To turn one off, just drag its slider from On to Off.

The Sounds setting screen also lets you specify Vibrate alerts by using two separate Vibrate sliders. When the Vibrate slider below the Silent heading is switched to On, the iPhone vibrates instead of making an audible alert whenever you turn off the ringer with the Ring/Silent switch on the side of the phone; otherwise, the phone issues no alerts at all when silenced. When the Vibrate slider below the Ring heading is switched On, the iPhone vibrates at the same time that it issues an audible alert.

Brightness

By default, the iPhone's display brightness is adjusted automatically, based on the light it senses around it. When you're outdoors on a sunny day, for example, the screen brightens; when you're inside a dark room, the display dims. If you'd like to override the automatic brightness settings—when you want to save battery power by making the display dimmer than the iPhone thinks necessary, for example—you do so in this screen. Turn auto-brightness off and drag the slider to adjust brightness up or down.

Wallpaper

On the iPhone, *wallpaper* refers to an image you choose, which appears when you unlock the phone or when you're talking to someone whose contact information doesn't have a custom photo attached.

To set and adjust your wallpaper picture, tap the Wallpaper control and then navigate to an image file in the collection provided by Apple (listed below the Wallpaper heading), pictures you've taken with the phone's camera, or the images you've synced to the iPhone. Just tap the image, and the iPhone will show you a preview of it as wallpaper (**Figure 2.17**).

Figure 2.17 Wallpaper sizing and preview.

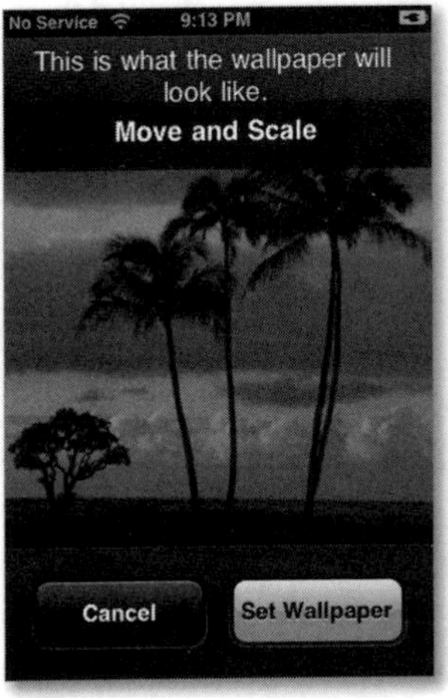

You can move any of your images by dragging them around or enlarge them by using the stretch gesture. When you're happy with the picture's orientation, tap Set Wallpaper. Interestingly enough, you can't resize any of the images that Apple includes in its Wallpaper collection (because, one assumes, they're perfect just as they are).

General

The General settings are ... well, pretty general. The grouping consists of a hodgepodge of miscellaneous controls (**Figure 2.18**).

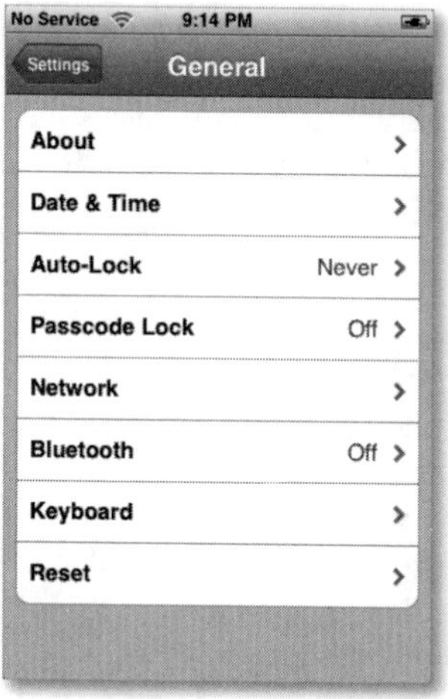

Figure 2.18 The General screen.

About

This screen provides your phone's vital statistics—the name of your network, the number of audio tracks, videos, and photos on the iPhone; total capacity; how much storage space remains; software version; serial and model numbers; Wi-Fi and Bluetooth addresses; the International Mobile Frequency Identity (IMEI) and Integrated Circuit Card Identifier (ICCID); the modem firmware number; and a Legal command that, when tapped, leads to a seemingly endless screen of legal mumbo jumbo.

Date & Time

The Date & Time settings include a 24- or 12-hour clock, a Set Automatically on/off switch (when it's set to on, the phone syncs to AT&T's clock; when it's set to off, you can enter a time zone, date, and time manually), and a Calendar area where you can set Time Zone Support on or off. As your iPhone tells you, "Time Zone Support always shows event dates and time in the time zone selected for calendars. When off, events will display according to the time zone of your current location."

Auto-Lock

The iPhone equivalent of a keypad lock, Auto-Lock automatically tells the touch screen to ignore taps after a customizable period of inactivity. Use these controls to specify that interval: 1, 2, 3, 4, or 5 minutes, or Never. To make the iPhone pay attention again, press the Home button.

When the iPhone is locked, it can still receive calls and SMS messages, and you can still use the iPhone's Volume switch to change the volume when listening to music or placing calls. The button on the headset's mic works when the iPhone is locked, too.

Passcode Lock

You'd hate to lose your iPhone. Worse, you'd hate to lose your iPhone and have some ne'er-do-well dig through it for your email, contacts, and schedule. If you fear that your iPhone could fall into the wrong hands (and yes, that may just mean your surly teenage daughter), create a passcode. To do so, tap Passcode Lock; then enter and re-enter a four-digit password with the numeric keypad (**Figure 2.19**).

Figure 2.19 The Passcode screen.

The next screen offers the option to turn the passcode off (useful if you decide that you no longer require a passcode), change it (for . . . well, you know), and a Require Passcode area that offers the options Immediately and After 1 Minute. There's also a Show SMS Preview option, which is on by default.

Network

The Network setting lists VPN and Wi-Fi networks. VPN (Virtual Private Network) is an encrypted network protocol used by many companies that allows authorized outsiders to join the company network, regardless of their location. When you choose VPN and then Settings, you're presented with a list of fields to fill in, including Server, Account, and Password. You can also choose between L2TP and PPTP networks.

When you tap the Wi-Fi entry in the Network screen, you're taken to the Wi-Fi Networks screen, atop which appears an on/off switch for enabling or disabling Wi-Fi on your iPhone. (Disabling Wi-Fi conserves power.) Below that is the Choose a Network area. Any Wi-Fi networks within range will appear in a list below. Those that have a lock icon next to them are password protected. To access a password-protected network, simply tap its name, enter the password with the keyboard that appears, and tap Join.

To see detailed network information, tap the blue › symbol to the right of the network's name. A new screen appears, listing such information as IP Address, Subnet Mask, Router, DNS, Search Domains, and Client ID.

tip **If a network that you never use routinely appears in this list, you can remove it by tapping its name and then tapping Forget This Network in the resulting screen.**

Finally, the bottom of the Wi-Fi Networks screen includes the Ask to Join Networks option. Leave this option set on (as it is by default), and your iPhone will automatically join known networks and ask to join a network if no known networks are available. If you switch the option off, you'll have to join networks manually without being asked. To do so, tap Other; then, using the keyboard that appears, enter the name of the network and password (if required).

Bluetooth

This setting is a simple on/off option. When you turn it on, the iPhone becomes discoverable and will search for other Bluetooth devices. Turning Bluetooth off can save power.

note **As this book goes to press, the iPhone's Bluetooth capabilities are pretty limited. The iPhone will pair with Bluetooth headsets, but you can't do certain things that you can do with many other mobile phones, such as copy files between the phone and your computer over a Bluetooth connection.**

Keyboard

Care to turn autocapitalization on or off (on means that the iPhone automatically capitalizes words after a period, question mark, or exclamation point)? Or to enable or disable Caps Lock (a feature that types in all capitals when you double-tap the keyboard's spacebar)? If so, this setting is for you.

Reset

If you'd like to remove information from your iPhone without syncing it with your computer, you use this screen, which includes a variety of options:

- **Reset All Settings.** This option resets your iPhone's preferences (your Network and Keyboard settings, for example) but doesn't delete media or data (your mail settings, bookmarks, or contacts, for example).

- **Erase All Content and Settings.** If your iPhone is packed with pirated music, and the Recording Industry Association of America is banging on the door, this option is the one to choose. It erases your preferences as well as removes data and media. After you're performed this action, you'll need to sync your iPhone with iTunes to put this material back on the iPhone.
- **Reset Keyboard Dictionary.** As you type on your iPhone's keyboard, word suggestions occasionally crop up. This feature is really handy when the iPhone guesses the word you're trying to type. If the word is correct, just tap the spacebar, and the word appears complete onscreen. But if the iPhone always guesses particular words incorrectly—your last name, for example—you can correct it by tapping the suggestion and continuing to type. The dictionary will learn that word.

  When you tap Reset Keyboard Dictionary, the iPhone's dictionary returns to its original state, and all additions—all you taught it—are erased.
- **Reset Network Settings.** To reset just the iPhone's Network settings, tap this entry.

note **Fear not that a slip of the finger is going to delete your valuable data. The iPhone always pops up a panel that asks you to confirm any Reset choice (Figure 2.20).**

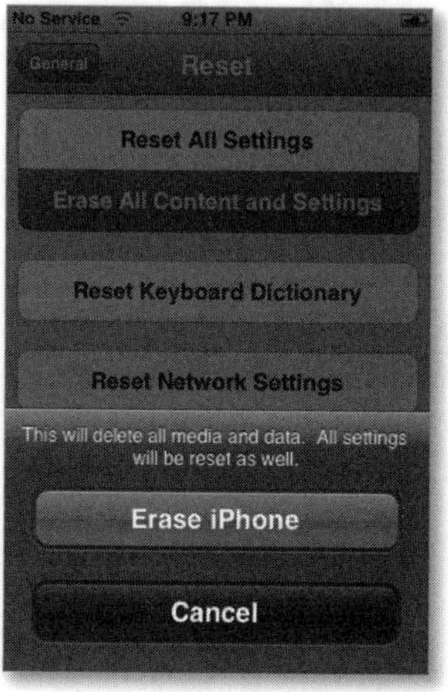

Figure 2.20 Confirming a reset in the Reset screen.

And more

The iPhone includes five more Settings screens: Mail, Phone, Safari, iPod, and Photos. Because these settings are intimately tied to specific iPhone functions, I discuss them in chapters devoted to those subjects.

3

Phone and SMS

Given the number of hats it wears—Internet communicator; music and video player; camera; picture viewer; personal information organizer; and ... oh, yeah, telephone—you can imagine Apple's marketing department scratching its collective head about the iPhone ad campaign.

"Best iPod ever?"

"The Internet in your pocket?"

"The mobile phone for the rest of us?"

"Room deodorizer?"

OK, maybe crossing the last one off the list was pretty easy. The point is, though, that the iPhone is far more than just a phone. But to be a success, it can't be anything less. It must be capable of making and answering calls as well as sending and receiving text messages. And it does all that—and more, as I show you in this chapter.

Calling All Callers

You've synced your contacts to your iPhone, and you're ready to make a call. The iPhone offers multiple ways to do it:

- **Do it the old-fashioned way.**

  Tap the Phone icon and then tap the Keypad icon at the bottom of the screen. On the keypad that appears, use the keys to tap out the number you want to call. Tap Call, and start talking (**Figure 3.1**).

- **Connect with Contacts.**

  Tap the Contacts icon; locate a contact; tap the contact's name; and in the resulting Info screen, tap the number you want to call.

Figure 3.1 The old-school keypad.

- **Revisit Recents.**

  If you recently had a phone conversation with someone, that person's number is likely in the Recents list. To find out, tap Recents and seek out the number. When you find it (the number or the contact associated with the number), tap it to place a call.

- **Favor Favorites.**

  If, while browsing through your phone, you added a person to the iPhone's Favorites list (the procedure for which I describe later in the chapter), tap Favorites and then tap that person's name. The iPhone will call him.

In-call options

When you place the iPhone against your face while making a call, its screen fades elegantly to black, but its advanced phone features remain at the ready. Pull the phone away from your face, and you'll see a series of option icons in the middle of the iPhone's screen (**Figure 3.2**).

Figure 3.2
In-call options.

The following sections explain these options.

Mute

If your spouse interrupts a call to ask who you're talking with, I advise you to tap this icon before issuing any reply along the lines of "That blowhard Charlie." Doing so turns the Mute icon blue and allows you to hear what the other party is saying but mutes the iPhone's microphone. To unmute the phone, just tap Mute again.

Keypad

Tap this icon to display a keypad after a call is in progress if you want to enter additional digits. This feature comes in handy for automated phone attendants that require you to enter account numbers, menu choices, and/or GPS coordinates before you can Speak To A Representative. To make the keypad disappear, tap Hide Keypad.

Speaker

The iPhone has speakerphone capabilities. To hear the call from the speaker, tap this icon; tap it again to listen to the iPhone's headset or receiver port.

Add Call

If you've ever tried to create a conference call on another phone, you know how complicated it can be. Not on the iPhone. The process works like this:

1. Tap the Add Call icon.

   The person you're speaking with will be put on hold. (You might warn her first that you're going to do this.)

2. Place another call.

   You can use the keypad (tap the Keypad icon to access it) or choose a contact (tap the Contact icon to view your contacts).

3. Tap Merge Calls (**Figure 3.3**).

   When that other caller connects, the Add Call icon turns into a Merge Calls icon. Tap this icon, and all three of you will be on the same call.

Figure 3.3 Tap the Merge Calls icon to create an instant conference call.

You can add more callers by repeating this procedure.

To boot someone from the call, tap the Conference icon that appears; tap the red Hang Up button next to the call; and then tap the End Call icon that appears.

If you'd like to commiserate privately with one of the other callers in the conference, tap Conference and then tap the Private button next to the caller (**Figure 3.4**). When you're ready to join the parties together again, tap Merge Calls.

Figure 3.4 While you're on a conference call, you can speak privately to one person or hang up on him.

If you'd like to add someone who is calling in to your conference, tap Hold Call + Answer and then tap the Merge Calls icon.

Hold

You know. Tap again to unhold.

Contacts

As I've pointed out before, this icon is helpful when you're using the Add Call feature. You can also browse your contacts while you're on a call.

tip **In addition to browsing your contacts while you're on a call, if you're connected to a Wi-Fi network (not EDGE), you can do pretty much anything other than use the iPod and YouTube functions. Check your stocks, look at the weather in Tasmania, tap out your grocery list, browse your photo collection, or use the Calculator to decide how much you're going to charge this client for taking your time. When you're ready to hang up or perform some other call-specific action, you can return to the call screen by tapping the green bar at the top of the iPhone's screen.**

Other buttons

Other buttons can appear during a call:

- **Ignore.** If a call comes in while you're on another call, and you'd rather send it to voice mail than speak with the person, tap the Ignore button that appears.
- **Hold Call + Answer.** To answer that incoming call and put the current caller on hold, tap Hold Call + Answer.

- **End Call + Answer.** For those "Whoops, that's the cheesemonger on the other line. Gotta go!" moments, tap End Call + Answer to drop the current call and answer the incoming call.
- **Swap.** You've put the Party of the First Part on hold to speak with the Party of the Second Part. To return to the PotFP and hold the PotSP, tap Swap, or tap the first caller's entry at the top of the screen.
- **Emergency Call.** I hope you never have to tap this button. The iPhone, like all mobile phones in the United States, can make emergency calls to special numbers (911, for example) when you're out of range of the network and even if your phone doesn't have a SIM card installed. But if you've locked your phone with a passcode and don't have time to unlock it, bring up the keypad, tap the Emergency Call button, and then tap out the emergency number.

Calling Plan

When the iPhone first hit the street, Apple placed the Phone application icon in prime position: in the first spot among the four major apps at the bottom of the Home screen. Tap that Phone icon, and you see one of the five Phone application screens: Favorites, Recents, Contacts, Keypad, or Voicemail. (Which screen appears depends on the last one you accessed before moving back to the Home screen or to another application.) A row of menu icons along the bottom of all five screens (refer to Figure 3.1) lets you navigate quickly among these screens. Because you may be in a hurry to place a call, I'll discuss them out of order.

Keypad

The function of this icon couldn't be much more obvious. Press Keypad, and you see ... a telephone keypad. To place a call, just tap the digits you want. As you tap, each digit appears in order at the top of the screen, nicely formatted with the area code in parentheses followed by the number—(555) 555-1212, for example.

In addition to the number keys, star (*), and pound (#), the keypad includes these icons:

- **Delete.** Tap the Delete icon to erase the last digit you entered. Tap and hold to delete a string of numbers quickly.
- **Call.** Tap Call to call the number you've entered.
- **Add Contact.** The Add Contact icon to the left of Call lets you create a contact quickly based on the number you've tapped in.

  Suppose that your dentist calls; he leaves a message that he needs a new boat and that your previously unmentioned impacted wisdom tooth will help him with the down payment. He asks you to call him back at 555-1234. You tap in the number, tap the Add Contact icon, and then choose Create New Contact if you have no contact for him or Add to Existing Contact if you don't have his new office number. (Or you can tap Cancel if you've thought better of the whole thing.)

Phone Facts

The iPhone is a Quad-band GSM (Global System for Mobile Communication) phone. The advantage of GSM phones is that compatible networks are widely available around the globe (particularly in Europe) and that they hold SIM cards, which you can conceivably move from one GSM phone to another.

The iPhone is a bit exceptional in this regard. Although the iPhone's SIM card will work in other phones, other phones' SIM cards currently don't work in the iPhone. But this iPhone-to-other phone scheme isn't so bad. If you had a lesser phone from AT&T/Cingular before purchasing your iPhone, you can carry that other phone as a spare. If your iPhone runs out of juice at a crucial moment, take out its SIM card, and put it in the other phone. The phone will work just as it once did.

Visual Voicemail

The iPhone offers a unique voice-mail system dubbed Visual Voicemail. What makes it different from other phones' systems is that you needn't wade through half a dozen messages to get to the one you really want to hear. Instead, all received messages appear in a list. You tap just the ones you want to listen to.

No one can be available 24 hours a day. Here's how to set up and use voice mail when you're not available to take a call:

1. Tap the Voicemail icon.

   When you first tap Voicemail, you'll be prompted to enter a password and record a voice greeting. When recording that greeting, it's not a bad idea to be somewhere quiet with good phone reception so that your greeting is as clear as possible.

   If you don't care to record a greeting, tap Voicemail and then tap Greeting. Tap the Default button, and callers will hear a canned greeting put together by AT&T.

note **To create a greeting at another time, just tap the Greeting button at the top of the Voicemail screen and then tap Custom. Tap Record; say your piece; then tap Done. Tap Play to listen to what you've recorded, and if you like it, tap Save.**

2. Locate a message you want to hear.

   Messages are named for the person who called (if known). The time (or date, if the call was made on a day other than the current one) appears next to the caller's name. If the caller is in the iPhone's list of contacts, a blue › icon appears next to her name. Tap that icon to be taken to her contact info screen. A blue dot marks each unheard message.

3. Select the message you want to listen to, and tap the Play icon on the left side of the message entry. The icon will change to a pause symbol..

   The iPhone downloads the message and then plays it (**Figure 3.5** on the next page).

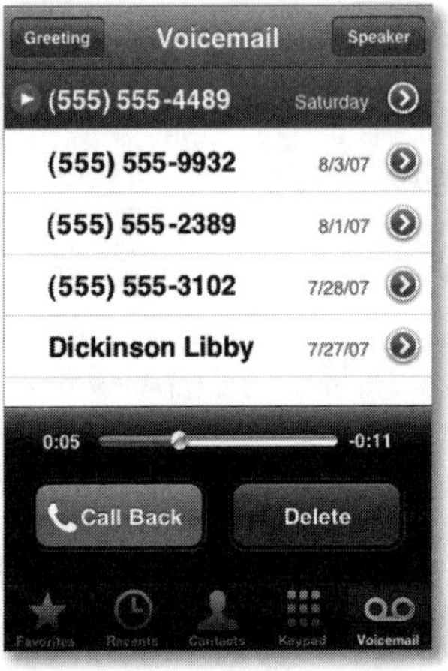

Figure 3.5 Playing a voice-mail message.

To pause a playing message, tap the Pause icon (the two vertical lines) to the left of the caller's name. Tap the Play symbol (the right-pointing triangle) to resume playing the message.

When you begin playing a message, a sheet appears that contains a scrubber bar and Call Back and Delete icons. To move through a message quickly—if your father tends to go on and on about the season's gopher issues before getting to the meat of the message, for example—drag the playhead to later in the message. If the phone was able to obtain a number through Caller ID, you have the option to call that person back immediately by tapping Call Back.

To listen to a message again, simply select it again and tap the Play icon. If listening once was enough, tap Delete.

note **Deleted messages aren't completely gone. At the end of your voice-mail list, you'll see a Deleted Messages entry. Tap this entry and then tap the message you'd like to listen to again. You can undelete a message by tapping it in the Deleted Messages screen and then tapping Undelete.**

4. Create a contact.

   As you read this chapter, you'll find that I talk a lot about contacts. The iPhone has no discrete Contacts application; the work of creating, adding, and managing contacts is handled largely by the Phone application.

   If someone who isn't in your list of contacts calls, and you'd like to add him, tap the blue > icon next to the message heading. In the sheet that appears, tap either Create New Contact (if this contact is new to you) or Add to Existing Contact (if the caller is already a contact, but you don't have this particular number). Then choose a contact to add the number to.

   You can also add the caller to your Favorites list, which is described in greater detail later in this chapter. Just tap the blue > icon and then tap Add to Favorites.

5. Listen later.

   The iPhone lets you know if you have voice-mail messages waiting. If you have missed one or more calls, received one or more voice-mail messages, or both, a red dot appears in the top-right corner of the Phone application icon in the Home screen. A number inside the dot denotes the combined number of missed calls and unheard messages. When you tap the Phone icon, a similar red dot appears over the Voicemail icon, indicating how many unheard messages you have.

Remote Control

Even though you own the coolest phone in the galaxy, you can do things with other phones that will enhance your iPhone experience, such as listen to your iPhone's messages from another phone. To do so, dial your iPhone's number, press star (*), enter your voice-mail password, and press pound (#).

This is more than a mere convenience. Checking Visual Voicemail from your iPhone counts against your monthly minutes. Call from a landline, however, and the call is free.

Similarly, you can record a greeting for your iPhone from another phone—not a bad idea, as a landline phone may produce a better-sounding message. Again, just dial your iPhone's number; press star; enter your password; press pound; and follow the canned voice's instructions to get to the greetings area, where you can record a greeting or an extended-absence message for your iPhone.

Recents

Like other modern mobile phones, the iPhone keeps track of calls you've made and received—both those you've participated in and those you've missed. You'll find a list of those most recent calls by tapping the Recents entry at the bottom of the iPhone's screen.

To see all calls, tap the All button at the top of the screen. Missed calls are shown in red (**Figure 3.6**). To see just your missed calls, tap the Missed button at the top of the screen. As in Voicemail, you see the time or day the call was made. Tap the blue › icon to be taken to one of a few screens, depending on what your iPhone "knows" about the phone number for each call you've placed or received.

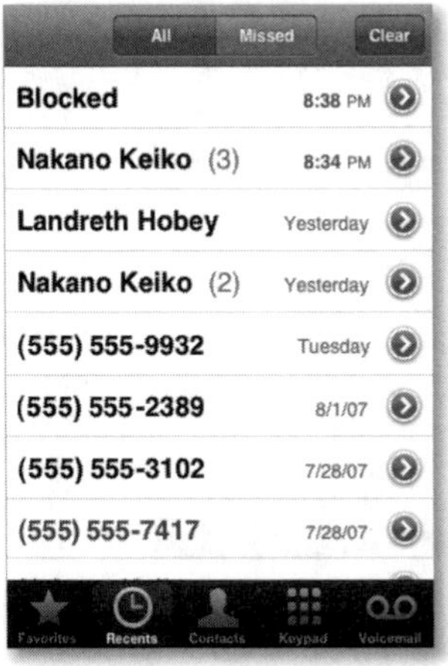

Figure 3.6 The Recents screen.

If a number belongs to someone listed in your contacts, tapping the › icon takes you to a screen that displays that person's contact info. At the bottom of that screen are icons marked Text Message and Add to Favorites. Tap Text Message to open a blank text message directed to that person. (If the person has more than one phone number in her contact info, a sheet will appear and ask you to choose the number to use.) Tap Add to Favorites, and that's just what happens: The person is added to your list of favorites, and you can access her info by tapping Favorites at the bottom of the screen.

If you place a call using a number that isn't in the iPhone's list of contacts, or if you capture an unrecognized Caller ID number from a received call, the phone number is displayed, and you can tap the blue › icon to view these options: Call, Text Message, Create New Contact, and Add to Existing Contact.

If the number came from a blocked number, the entry will read *Blocked Caller*, and tapping the › icon will tell you only the date and time the person called. Tap Clear to clear the Recents lists.

Favorites

Use the Favorites list to store those very special contacts you call routinely (**Figure 3.7**). Here, you'll find the numbers you've added by tapping the

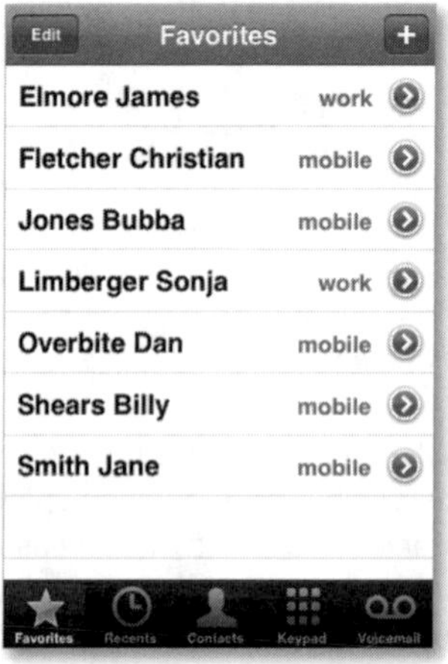

Figure 3.7 The Favorites screen.

Add to Favorites button in a contact's Info screen or by tapping Favorites, tapping the plus (+) icon, and then navigating through your list of contacts to find a name. Only those contacts that contain phone numbers are eligible to be added to Favorites. Tap a contact that doesn't have a phone number, and you'll see the No Phone Numbers message, which is the iPhone's way of saying, "Sorry, Charlie."

tip

Calling and driving is dangerous, so I never initiate a call unless I'm stopped. But by *stopped*, I could mean sitting at a stoplight and needing to get to a number in a hurry. To stay safer, I add contacts to my list of favorites that I'm likely to call when driving—a work-at-home co-worker whom I need to alert when I'm planning to drop by, or my wife when we're coordinating school pickups for our child.

Just like those in Recents and Voicemail, each entry in the Favorites list bears a small blue > icon. Tap it to view that person's Info screen.

This may sound silly, but some people care very much where they rate in your Favorites list. If your husband or mother isn't at the top of the list and should be, here's how to put things right: While in the Favorites screen, tap Edit; then drag contacts around, using the bars to the right of each name. If someone no longer rates enough to be a favorite, tap the red minus (–) icon next to his name to reveal the Remove button. Tap Remove, and that person will be removed from Favorites (but not from your contacts).

Contacts

Last but hardly least is Contacts, the Big Kahuna of the Phone applications—a Kahuna that makes its presence known in just about every iPhone application save the iPod area. Although you'll find it more convenient to ask your address-book application to do the heavy lifting in regard to creating and editing contacts (because it's far easier to enter information from a real keyboard than on the iPhone's virtual keyboard), you can do a lot of cool things with contacts directly on your iPhone.

The people you know

Tap the Contacts icon, and you see a list of your contacts in alphabetical order (**Figure 3.8**). As a bonus, you see your iPhone's number at the top of the screen. (You'd be amazed how many people

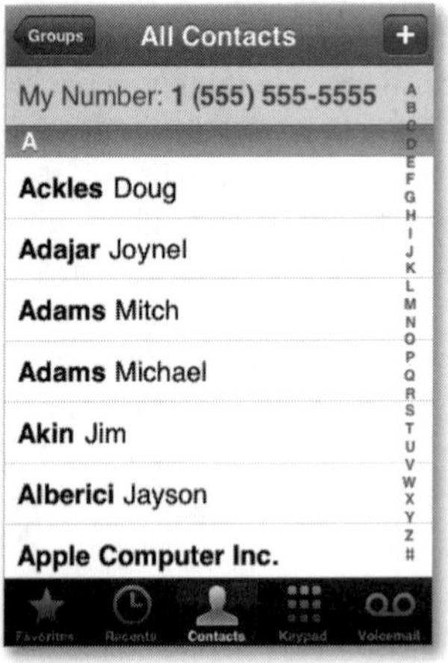

Figure 3.8 The Contacts screen.

don't know their mobile number, and this feature is a great reminder.)

This list works very much like any long list of items you see in the iPhone's iPod area. A tiny alphabet runs down the right side of the screen. Tap a letter to move immediately to contacts whose names (first or last, depending on how you've configured name sorting in Phone preferences) begin with this letter.

When you tap a name, you're taken to that contact's Info screen (**Figure 3.9**).

Figure 3.9
A contact's Info screen.

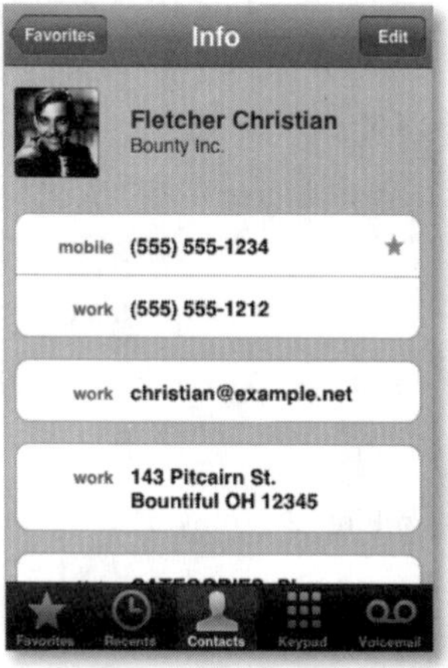

Here, you can find information including the following:

- Photo

  This item can be a photo you've added in Address Book on a Mac, by tapping Add Photo and choosing a picture from your Photos collection, or by assigning a picture to a contact in the Photos or Camera application.

- Name
- Title
- Company
- Phone number

  Possible phone headings include Mobile, Home, Work, Main, Home Fax, Work Fax, Pager, Home Phone 2, Work Phone 2, and Other. There's even a Custom Label option so you can enter labels such as *Dirigible* or *Private Train Car*.

- Email address

  This item includes Home, Work, and Other options.

- URL (for the contact's Web site)
- Home address
- Work address
- Other address
- Birthday
- Notes
- Other fields

  Your options include Prefix, Middle Name, Suffix, Nickname, Department, and Anniversary Date.

You won't necessarily find all these entries in a contact's Info screen; this list just shows you what's possible to include.

Organizing contacts in groups

Although you see a list of all your contacts when you first tap Contacts, the Contacts application has an organizational layer above the main list. If, in the Info iPhone preference within iTunes, you've chosen to sync your address book with select groups of contacts, or if your full address book contains groups of contacts, those groups will appear in the Groups screen, which you access by tapping the Groups button in the top-left corner of the Contacts screen (**Figure 3.10**).

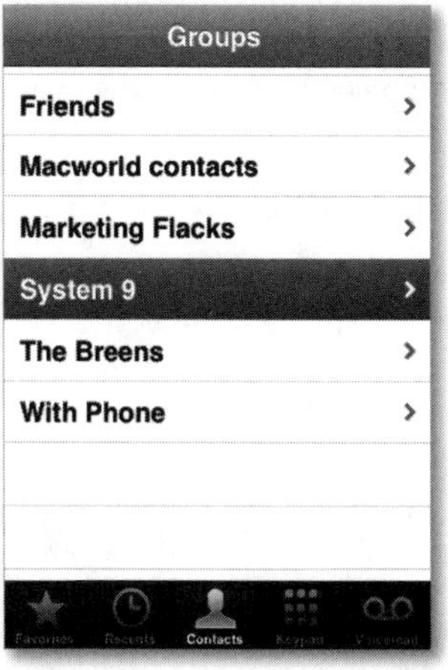

Figure 3.10 The Groups screen.

Organizing in groups makes a lot of sense if you have loads of contacts. Although Apple made traversing a long list of contacts as easy as possible, easier still is tapping something like a Family group and picking Uncle Bud's name out of a list of 17 beloved relatives.

tip I've found it really helpful to create a group that includes just contacts that have phone numbers. The iPhone will sync all your contacts, whether or not they include phone numbers, but more than anything else, I need phone numbers on my phone. In Apple's Address Book, you can create just such a group easily. Choose File > New Smart Group, and configure the resulting pane to read *Phone Is Set*. This step places all contacts that contain a phone number in their own group. Then you can sync this group to your iPhone so that you don't have to bother trying to call contacts you don't have phone numbers for. Regrettably, iPhone-compatible Windows applications (Outlook, Outlook Express, and Windows' Address Book) with address-book functionality don't have this kind of easy-does-it feature.

Making contacts

The best way to become familiar with the iPhone's contacts is to make some of your own. To do that now, tap Contacts and then the plus icon in the top-right corner of the iPhone's screen.

Anatomy of the New Contact screen

The New Contact screen contains fields for the elements I listed in "Organizing contacts into groups" earlier in this chapter—Photo, First and Last Name, Phone Numbers, Email Addresses, URL, Physical Addresses (Home, Work, and Other, for example—as well as an Add Field entry (**Figure 3.11**). To add information to one of these fields, tap the field or the green plus icon to its left. In the resulting screen, you'll find a place to enter the information.

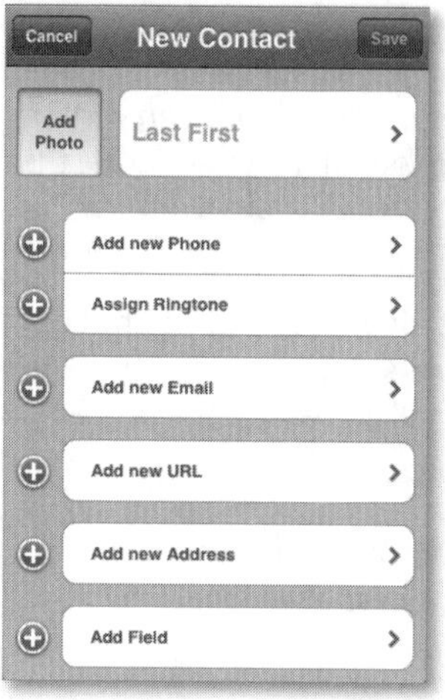

Figure 3.11 The New Contact screen.

Here are the special features of each screen:

- **Add Photo**

  Tap this entry to display a sheet containing buttons marked Take Photo, Choose Existing Photo, and Cancel. Tap Take Photo, point it at the object you'd like to capture (ideally, something more pleasing than the back end of your cat), and tap the green camera icon. If you don't like what you've taken, tap Retake, and do just that. You can resize and reposition the picture by pinching the image and dragging it around the screen. When you settle on a view you like, tap Set Photo.

  You can edit this picture later, if you like. Tap Edit, and you get the options Take Photo, Choose Existing Photo, Edit Photo (which takes you to the Move and Scale screen, where you can pinch and reposition), and Delete Photo.

When you tap Choose Existing Photo, you're taken to your Photos library, where you can select a picture. Just as you can with the pictures you take, you can move and scale these images and then tap Set Photo to attach them to the contact.

And why all the fuss about a contact's picture when it appears in this tiny box? The photo appears across much of the iPhone's screen when you talk to that person on your iPhone.

- **Name**

  In this screen, you enter first, last, and company names. Tap Save to return to the New Contact screen.

- **Add New Phone**

  As the name says, this field is where you add a phone number. In the Edit Phone screen, you tap in the number from the keypad and then choose the kind of phone number: Mobile, Home, Work, Main, Fax, Pager, or Other.

note **The numeric keypad contains a key that reads +*#. Tap it, and these three characters appear on the keypad's bottom three keys. What good are they? They're used by automated answering systems for performing certain functions. Some phone systems, for example, require that you press the pound key and then a key combination to unblock a hidden phone number or append an extension. Another character can introduce a pause between numbers.**

- **Assign Ringtone**

  Tap this field, and you can assign one of 25 ringtones to the contact. This feature is a great way to know who's calling without having to pull the iPhone out of your pocket.

- **Add New Email**

  Enter your contact's email addresses here. The iPhone's keyboard in this screen contains @, period (.), and .com keys to make the process easier.

- **Add New URL**

  Same idea here. The more-convenient keyboard is in evidence, but instead of an @ symbol, you'll find period (.), slash (/), and .com. You can apply a Home Page, Home, Work, or Other label to the URL.

- **Add New Address**

  In the United States, the default Edit Address screen contains two Street fields and areas for City, State, and Zip. Ah, but tap the country field and choose a different nation from the list that appears, and these fields change. If you choose Ukraine, for example, the bottom fields change to PostalDistrict, Province, and PostalCode. Tap the Location icon next to the Country icon to choose the nature of this address: Home, Work, or Other.

- **Add Field**

  Tap Add Field, and you can add more fields to a contact's Info screen. These fields include Prefix, Middle Name, Suffix, Nickname, Department,

Birthday, Date (Anniversary and Other are the options), and Note. Both the Birthday and Date screens contain the iPhone's spinning date wheel for selecting month, day, and year quickly.

Working with existing contacts

Once you have contacts on your iPhone, you can delete them, edit the information they contain, or use that information to perform other tasks on your iPhone.

To delete a contact, just tap the Edit icon that appears in the contact's Info screen, scroll to the bottom of the screen, and tap the big red Delete Contact button. You'll be asked to confirm your choice.

To edit a contact, tap that same Edit icon in the contact's Info screen, and make the edits you want (**Figure 3.12**). You can add information by tapping a field that begins with the word *Add* (or just tap its green plus icon). To delete information, tap the red minus icon next to the information and then tap the now-revealed Delete button. When you're finished editing the contact, tap Done.

As for initiating actions on your iPhone via a contact's Info screen, most of the elements in the screen are *live,* meaning that if you tap them, something happens. If you tap a phone number in the Info screen, for example, the iPhone dials that number; tap an email address, and a New Message window appears in the Mail application, addressed to that person. If you tap a URL, Safari opens and takes you

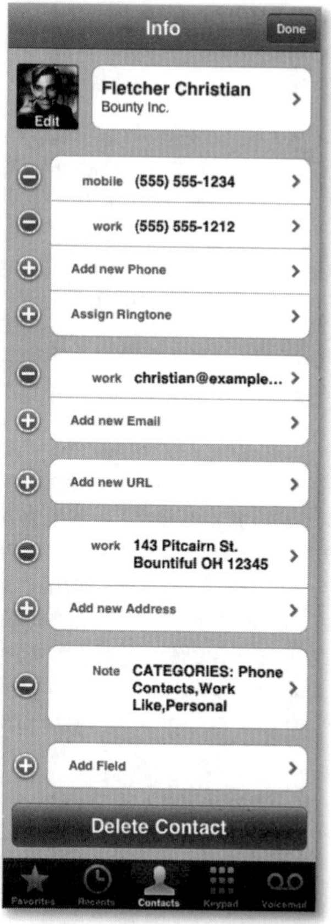

Figure 3.12 An elongated view of the contact edit screen.

to that Web page. Tap an address, and Maps opens to show you its location.

At the bottom of an Info screen that contains at least one phone number, you'll find Text Message and Add to Favorites buttons. Because each of these functions requires a phone number, you can see why they're present here but not in Info screens that don't bear phone numbers.

You already know about Favorites. When you tap Text Message, and the contact has more than one phone number in the Info screen, a sheet rolls up that contains each phone number. Tap the phone number you'd like to use, and SMS opens with the contact's name at the top of the screen.

And gosh, speaking of SMS ...

Text

When Steve Jobs first demoed the iPhone, it appeared to contain an instant-messaging client similar to Apple's iChat. That turned out not to be the case. The iPhone's SMS (Short Message Service) application, Text, looks very much like iChat, but it's not. It's a standard text-messaging service much like the ones you find on other mobile phones.

Sending messages

Using Text is pretty straightforward. Tap Text in the iPhone's Home screen, and you'll see a Text Message screen. Tap the New Message icon in the top-right corner of the screen to compose a new message.

In the middle of the resulting New Message screen, you'll see the text field with a To field for your recipient's name or phone number above and the iPhone's alphabetic keyboard below.

To enter a name in the To field, tap it and then begin typing the name of the person you want to message. As you type, a list of matching names from your contacts appears (**Figure 3.13**). Continue to type, and the list narrows. To select a recipient, just tap that person's name.

Figure 3.13 Start typing a name in Text, and your iPhone will suggest recipients.

Alternatively, you can tap the plus icon in the To field, which brings up the iPhone's Contacts list. Navigate to the contact you want to contact, and tap that person's name to add it to the To field.

tip **Why type a name when you can choose a contact? The Contacts screens include *all* your contacts—even those that don't contain phone numbers. When you begin typing on the alphabetical keyboard, only those contacts with phone numbers appear in the list.**

Finally, you can simply enter a phone number. Tap the 123 icon at the bottom of the alphabetical keyboard to produce the numeric keypad. Use the keypad to enter the number of the person you want to call.

Tap in the text field, and start typing. The field will show four lines of text. When you're ready to send your message, tap Send (**Figure 3.14**).

Figure 3.14 An SMS message session in progress.

Receiving messages

When it receives an SMS message, the iPhone alerts you—by sounding the New Text Message alert configured in the Sounds setting or (if you've switched the iPhone to silent mode by toggling the Ring/Silent switch) by vibrating. The iPhone will also display a preview of the first part of a received message (**Figure 3.15**). When you look at the Home screen, you'll see a red circle with a number inside denoting the number of unread text messages on your phone.

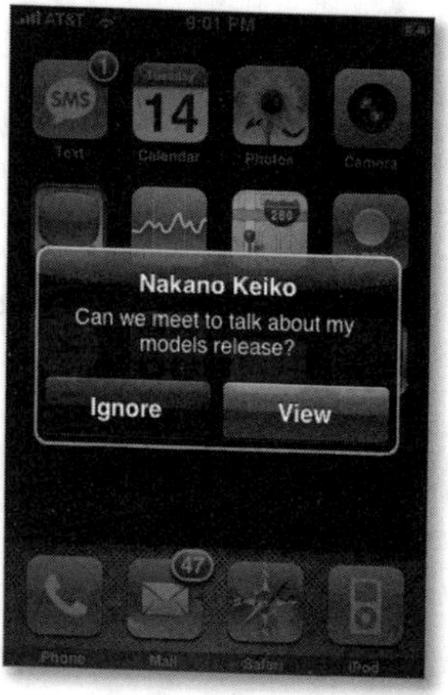

Figure 3.15 Your iPhone will display a preview of received SMS messages unless you tell it not to.

Private Previews

Having the first bit of an SMS message displayed on your phone's face isn't always ideal. Some very personal stuff can come across via SMS—personal stuff that you might prefer that your mother not see when she "accidentally" picks up your iPhone. To prevent this situation, you need to create a passcode for your iPhone. To do so, tap Settings and then tap General. Tap Passcode Lock, and create a passcode. After entering and confirming a passcode, you'll see a Show SMS Preview option at the bottom of the Passcode Lock screen. Toggle the switch to Off, and the SMS preview won't appear.

To view your messages, just tap Text. Any received messages will appear in the Text Messages screen (**Figure 3.16**). Unread messages are marked by blue dots. To read a message, just tap it. To reply to that message, enter the text in the Send field with the iPhone's keyboard, and tap Send. Similarly, you can tap any message in the list and send a new message to that person simply by entering text and tapping Send. You can access that person's contact information by tapping the Contact Info button or place a voice call by tapping Call.

Figure 3.16 Tap an SMS message to view the course of your conversation.

To delete entries from the Text Messages list, just tap the Edit button; then tap the red minus icon, and tap Delete. The swipe-and-delete trick works here, too. Just swipe your finger to the left or right across the message entry, and tap the Delete button that appears.

Text messages can also contain live links. If someone places a phone number in a text message, for example, you can tap it, and the iPhone will call that number. Email addresses, URLs, and physical addresses are live, too. Tap an email address, and Mail opens with a message addressed to that person. Tap a URL, and Safari launches and takes you to that site. And tap a street address, and Maps opens to reveal that location on a Google map.

The Limits of Text

If you've used SMS on other phones, you may be surprised by the limitations of the iPhone's Text application. Allow me to cushion the blow by revealing those limitations here:

- You can't send SMS messages to a group. One message. One recipient. Period.
- You can't forward an SMS message, meaning that you can't copy a message and then send it to another person.
- You can't send media files via SMS. This capability, called MMS (Multimedia Message Service), is not supported by the iPhone.
- You don't get character-count warnings. By definition, an SMS message is up to 160 characters long, and that's what AT&T charges you for. Text is happy to let you type more than 160 characters in a single message, though, and it won't warn you that you're being more verbose than you may desire. So if money is an object, keep your text messages short.

4

Mail and Calendar

Seeking a less-immediate way to communicate than the phone or SMS? Can't figure out how to copy your notes and photos from your iPhone to a computer not synced with your iPhone, or how to receive documents that you can view on your iPhone? Or is your life so tied to email that you can't stand to be away from your computer for more than a couple of hours? If so, you and the iPhone's Mail application are about to become best friends.

Portable email is a real boon, and so is knowing where you're supposed to be from one minute to the next. To help with the latter, the iPhone includes a Calendar application that lets you sync your schedule with your Mac or Windows PC, as well as create calendar events on the go. In this chapter, I explain the ins and outs of both applications.

Using Mail

Mail is a real email client, much like the one you use with your computer. With it, you can send and receive email messages, as well as send and receive a limited variety of email attachments. You can send photos, for example, and receive JPEG graphics files, text, HTML, Microsoft Word and Excel documents, and Adobe PDF documents. Regrettably, you can't edit any of the files you receive; they're read-only.

tip

The iPhone can read these document extensions: .c, .cpp, .diff, .doc, .docx, .h, .hpp, .htm, .html, .m, .mm, .patch, .pdf, .txt, .xls, and .xlsx. For an attachment to be readable, the original file must bear one of these extensions. If you receive an email message with the Word attachment Breen's Hot iPhone Tips!, for example, you won't be able to read it on the iPhone unless it's named Breen's Hot iPhone Tips!.doc.

Mail is limited in some other ways:

- It has no option for sending a bcc (blind carbon copy) to anyone but yourself. (You'd use a bcc to send someone a copy of a message without letting the recipients in the To field know you've done so.)
- Unlike all modern computer-based email clients, the iPhone has no spam filter and no feature for managing mailing lists.

- You can't flag messages or apply rules that allow Mail to sort or copy certain messages (those from a particular sender, for example) into specific mailboxes.
- Speaking of mailboxes, you can't create new mailboxes on the iPhone, either. Instead, you must create them on your computer or the Web, and you can do so only with IMAP accounts; they'll appear in Mail after you sync the mail accounts on your computer with the iPhone.

The iPhone is capable of sending and receiving email over a Wi-Fi connection and AT&T's EDGE network. Other than the speed of sending and receiving messages, there's no significant difference between running Mail over either network. Note, however, that there's a big difference if you're using your phone overseas. Wi-Fi costs nothing extra, but AT&T imposes punitive roaming charges for using EDGE (for email or anything else) outside the United States.

Now that you know what Mail can and can't do, you're ready to look at how to use it.

Account creation

When you first synced your iPhone to your computer, you were asked whether you wanted to synchronize your email accounts to the phone. If you chose to do so, your iPhone is nearly ready to send and receive messages. All you may have to do now is enter a password for your email account.

But I'm getting ahead of myself. Rather than start in the middle, with a nearly configured account, I'll start at the beginning so that you can follow the iPhone's account-setup procedure from start to finish. Here are the steps you take:

1. Tap the Settings icon in the iPhone's Home screen.
2. Tap Mail.
3. Tap Add Account.

   You'll see a page that offers Yahoo Mail, Gmail, .Mac, AOL, and Other. The first four options are Web-based email services; see the sidebar "Configuring Yahoo Mail, Gmail, .Mac, and AOL" in this section for details.

4. Tap Other.

   I ask you to tap Other because this option lets you set up email accounts that aren't among the four listed in step 3. In the resulting screen, you can configure IMAP, POP, and Exchange accounts. The configuration options for each kind of account are the same.

5. Tap Name, and enter your real name (as opposed to your user name).
6. Tap Address, and enter your email address (such as example@examplemail.com).
7. Tap Description, and enter a description of your account.

   I often use the name of my account for this entry—Macworld, for example.

Configuring Yahoo Mail, Gmail, .Mac, and AOL

The iPhone's designers made configuring one of these accounts really easy. Rather than forcing you to enter arcane server settings, the iPhone understands what those settings are and takes care of that part of the business for you. Instead, when you tap any of these items, you'll be asked only for your real name, email address on the service (example@gmail.com, for example), the account password, and a descriptive name for the account.

8. Below Incoming Mail Server, tap Host Name, and enter the name of the server for incoming mail.

   This information, provided by your Internet Service Provider (ISP), is in the format mail.examplemail.com.

9. Tap User Name, and enter the name that precedes the at (@) symbol in your email address.

   If the address is bruno@examplemail.com, for example, type **bruno**.

10. Tap Password, and enter the password for your email account.

note **Type the password very carefully. Like other iPhone password fields, this one doesn't display characters you type—just black dots—so you can't check your work when you're done.**

11. Below Outgoing Mail Server (SMTP), tap Host Name, then enter the appropriate text, which, once again, will be provided by your ISP, typically

in the form of smtp.exampleemail.com. Tap the User Name and Password fields and enter text as you did in steps 9 and 10 above.

tip

If your ISP's SMTP server uses a port other than the standard port 25 (your ISP will tell you which port it uses), you'll have to append a colon and then the port number to the end of the SMTP host-name address, as in smtp.examplemail.com:587.

12. When you've double-checked that everything is correct, tap Save in the top-right corner of the screen.

The account appears in the list of accounts in the Mail Settings screen (**Figure 4.1**).

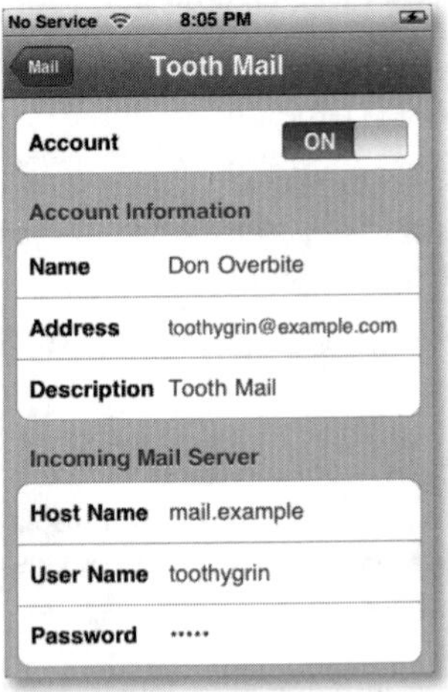

Figure 4.1 Configured email POP account.

Further configuration

Most people can stop right here and get on with mucking with Mail, but your email account may require a little extra tweaking for it to work. Here's how to do just that:

1. Tap your account name in the Mail Settings screen.
2. If you'd like that account to appear in Mail's Accounts list, be sure that the Account slider is set to On.

   Why turn it off? Perhaps you've got a load of messages sitting on the server that you'd rather not download with your iPhone. Download those messages with your computer, delete them from the server, and then enable the account on your iPhone.
3. Verify that the information in the account's settings fields is correct; if not, tap in the field you want to edit and start typing.
4. Tap the red Advanced button at the bottom of the screen, and choose the options you want in the resulting Advanced screen for POP accounts (**Figure 4.2** on the next page). Use these settings to specify:
   - When you want deleted messages to be removed from the iPhone
   - Whether your account will use Secure Sockets Layer (SSL) protection to transmit and receive email

- The kind of authentication your account requires (MD5 Challenge-Response, NTLM, or Password)
- When you want email to be deleted from the server (options include Never, Seven Days, and When Removed from Inbox).

This information is individual enough that I'll leave it to your IT or ISP representative to tell you how to configure these options. Worth noting, however, is that you may be able to suss out these settings by looking at how the email client on your computer is configured.

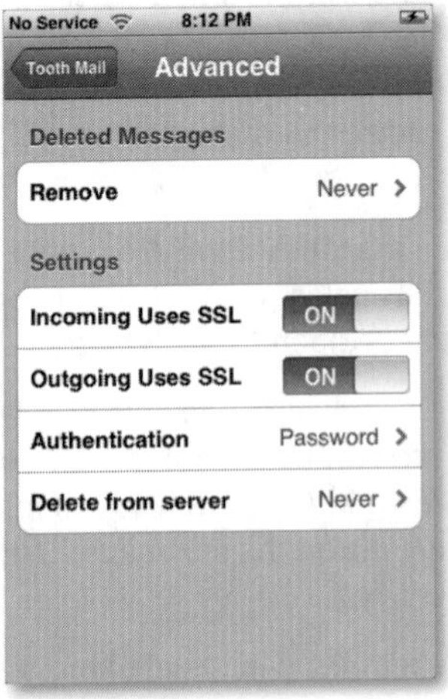

Figure 4.2 A POP account's Advanced settings.

For IMAP accounts, you have some different options. You can choose which mailboxes will hold drafts, sent email, and deleted messages. You can choose when to remove deleted messages (Never, After One Day, After One Week, After One Month). You can also turn on or off Incoming and Outgoing SSL (note that Yahoo Mail doesn't offer an SSL option). You can choose the same authentication schemes as your POP-using sisters and brothers. And you can enter an IMAP path prefix—a pathname required by some IMAP servers so that they can show folders properly.

tip **If you're like me, you want to have all your email in one place, and that one place is on your computer. If you receive the same email on both your computer and your iPhone, you risk having some of it here and some of it there if you ask both your computer and iPhone to delete it automatically. If this describes your setup, configure the iPhone's email accounts so that they never delete email from the server. On your computer, configure server settings so that email is either never deleted (meaning that you'll do it manually) or deleted very rarely so that you'll have the opportunity to retrieve it on your iPhone, if you like.**

Mail behavior

Before I leave the Mail Settings screen, I should examine the options that tell Mail how to behave (**Figure 4.3** on the next page).

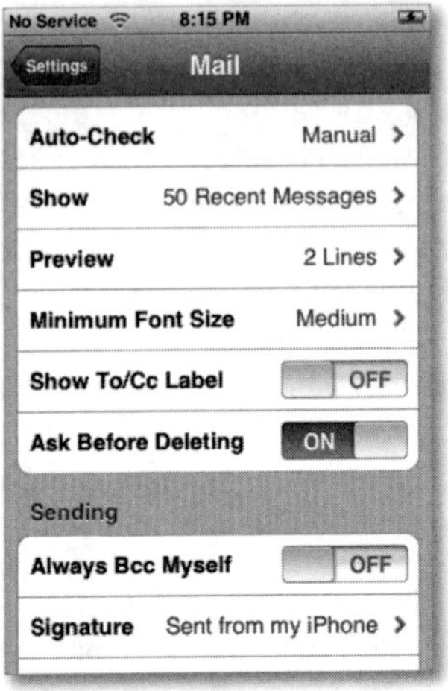

Figure 4.3 Additional Mail settings.

View the bottom part of the screen, and you find these options:

- **Auto-Check.** How often would you like Mail to check for new messages? Options include manual (Mail checks when you launch the Mail application and then only after you tap the Retrieve icon) and every 15, 30, or 60 minutes.
- **Show.** How many messages would you like Mail to display? Options include 25, 50, 75, 100, or 200 recent messages.
- **Preview.** When you view message subjects within a mailbox in one of your Mail accounts, you see the first bit of text in each message. The Preview entry determines how many lines of this text you'll see: none, 1, 2, 3, 4, or 5 lines.

- **Minimum Font Size.** This setting determines how large the text will be in your email messages: Small, Medium, Large, Extra Large, or Giant. Medium is good for most eyes, and it saves a lot of scrolling.
- **Show To/Cc Label.** When this option is set to on, Mail plasters a *To* next to messages that were sent directly to you and a *Cc* next to messages on which you were copied.
- **Ask Before Deleting.** When you set this option to on, if you tap the Trash icon to delete the message you're reading, you'll be asked to confirm your decision. If you swipe a message and then tap the red Delete icon that appears, however, you won't be asked for confirmation.
- **Always Bcc Myself.** If you're the kind of person who wants a copy of every message you send (but don't want the recipients of those messages to know), switch on this option. You'll receive a copy of each message you send.
- **Signature.** Ever wonder where that proud *Sent from My iPhone* message comes from—the one that appears at the bottom of every message you send from your iPhone? Right here. As a new iPhone owner, you'll want to stick with this default message for a while, simply for the bragging rights. Feel free to tap this option later and enter some pithy sign off of your own.
- **Default Account.** Although this option is the last to appear in the Mail Settings screen (it's so far down the screen that Figure 4.3 doesn't show it),

it's one of the most important settings. If you have more than one email account set up, it determines which account will send photos, notes, and YouTube links. When you send one of these items, you can't choose which account sends it, so give this option some thought. You may discover that Wi-Fi hotspots are reluctant to send mail through your regular ISP's SMTP server, whereas Gmail accounts rarely have this problem. For this reason, you may want to make your Gmail account the default.

Sending and Receiving Mail

Now that your accounts are *finally* set up properly, you can send and receive messages. The process works this way.

Receiving email

Receiving email is dead simple. Just follow these steps:

1. Tap the Mail icon in the iPhone's Home screen.

   Even if you have the Auto-Check Mail Setting set to Manual, Mail will check for new messages when you first launch the application. If you have new messages, the iPhone will download them.

   When it does, a number will appear next to the account name, indicating the account's number of unread messages. Any messages that contain attachments will bear a paper-clip icon next to the sender's name.

2. Tap the account name.

   You'll see a list of that account's mailboxes (**Figure 4.4**). For POP accounts, those mailboxes will include Inbox, Drafts (if you've saved any composed messages without sending them), Sent (if you've sent any messages from that account), and Trash (if you've deleted any messages from that account). For IMAP accounts, you'll see Inbox, Drafts, Sent, Trash, and then any folders associated with that IMAP account (folders you've added to a .Mac account, for example).

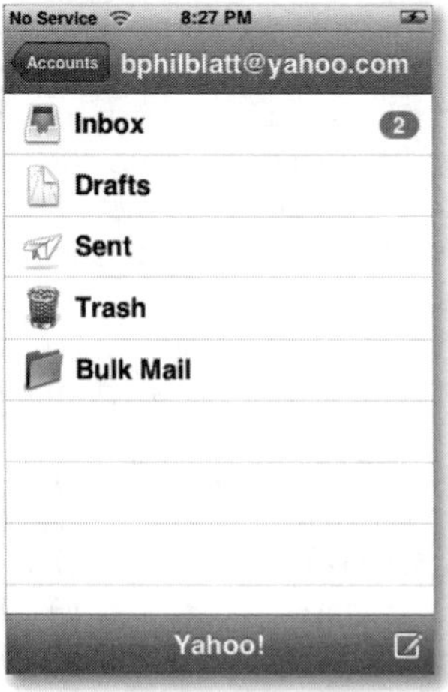

Figure 4.4 An account screen.

   In the bottom-right corner of an account screen, you'll see a Compose icon. Tap it, and a New

Message screen appears, along with the iPhone's keyboard. I talk about creating new messages in "Creating and sending email" later in this chapter.

3. Tap the Inbox.

   Messages will appear in a list, with the most recently received messages at the top. Unread messages will have a blue dot next to them. The Inbox heading will have a number in parentheses next to it—*Inbox (22)*, for example. That *(22)* means that you have 22 unread messages.

   This screen also bears a Compose icon and, in the bottom-left corner, a Retrieve icon, which you tap to check for new mail.

   This screen also has an Edit button in the top-right corner; this icon lets you delete messages. Tap it, and all the messages in the list acquire a red minus (–) button. Tap a minus button, and a Delete button appears to the right of the message. Tap Delete to move the message to the Trash. Alternatively, you can do without the Edit button by swiping your finger across a message entry to force the Delete button to appear. Again, tap Delete, and the message moves to the Trash.

note **When you delete a message, it's not really gone; it's simply moved to the Trash mailbox. To delete the message for real, tap Trash. Then swipe the message and press Delete; or tap Edit, tap the minus icon, and tap Delete. (Yes, I agree that this process is tiresome.) Then—and only then—is the message truly gone. Oh, and as we go to press, the iPhone lacks Mark All As Read and Delete All commands. That makes me sad.**

Spam and the iPhone

As I write these words, the iPhone's Mail application lacks a spam filter—a utility that looks through your incoming email for junk mail and quarantines it in a special mailbox. This lack is a drag if you're using an account that attracts a lot of spam.

My solution? Don't use such an account on your iPhone. Google offers its free Gmail email service at www.gmail.com. Gmail provides loads of email storage (probably more than your ISP), and you can access it from the Web, your iPhone, and your computer's email client. Best of all, it offers excellent spam filtering. Like I said, it's free. Give it a try.

Navigating the Message screen

Simple though it may be, the Message screen packs a punch. In it, you find not only standard email elements such as From and To fields, Subject, and message body, but also icons for adding contacts and for filing, trashing, replying to, and forwarding messages. The screen breaks down this way.

Before the body

The top of the Message screen displays the number of messages in the mailbox as well as the number of the displayed message—*2 of 25*, for example. Tap the up or down arrow to the right to move quickly to the previous or next message in the mailbox (**Figure 4.5** on the next page).

Figure 4.5 Message body with document attached.

Below that, you'll see From and To fields. Each field will display at least one contact name or email address (one of which could be your own) in a blue bubble. Tap one of these bubbles, and if the name or address is in your iPhone's Contacts directory, you'll be taken to its owner's Info screen. If the name or address isn't among your contacts, a screen will appear, offering you the option of emailing the person, adding him to your Contacts directory, or adding the address to an existing Contacts entry.

Tapping Email with a contact bubble selected opens a new email message with that person's email address in the To field. The email will be sent from the account you're currently working in.

Tap Create New Contact, and a New Contact screen appears, with that person's name at the top and his email address filled in below. If the message has no name associated with it—you were sent a message by a company address such as info@example.com, for example—no name will appear in the Name field (**Figure 4.6**).

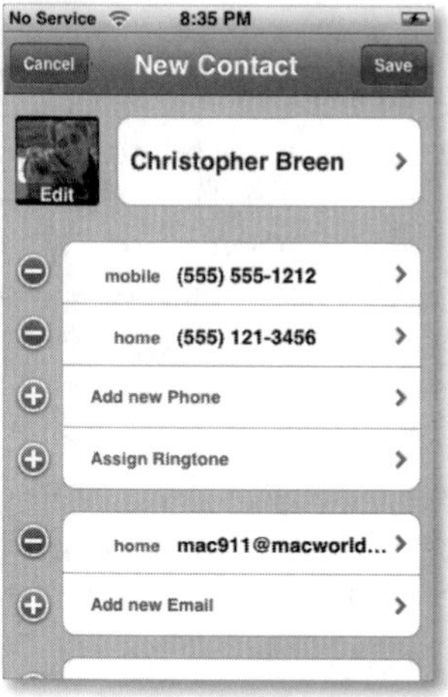

Figure 4.6 Create a new contact.

Tap Add to Existing Contact, and a list of all the contacts on your iPhone appears. Tap a contact, and an Add Email screen appears, displaying that email address in the first field. Below that field is another that reads Home, Work, or Other. Tap it, and assign an appropriate label to the contact.

You can hide the To field by tapping the Hide entry next to it. This action will hide all the To fields in all the messages in all your accounts, and it will make the Hide entry change to read Details. To expose the To fields again, just tap Details.

Below the From and To fields, you'll see the message subject, followed by the date and a Mark As Unread entry. Tap this entry to do exactly what it suggests.

Body talk

Finally, in the area below, are the pithy words you've been waiting for. Just as in your computer's email client, you'll see the text of the message. Quoted text appears with a vertical line to its left—or more than one line, depending on how many quote "layers" the message has. If a message has several quote layers, each vertical line is a different color. (The first three layers are blue, green, and red, respectively; subsequent layers are red from there on out.)

If the message has attachments, they will appear below the message text. If Cousin Bill sends photos from his latest vacation, they'll appear here (**Figure 4.7**).

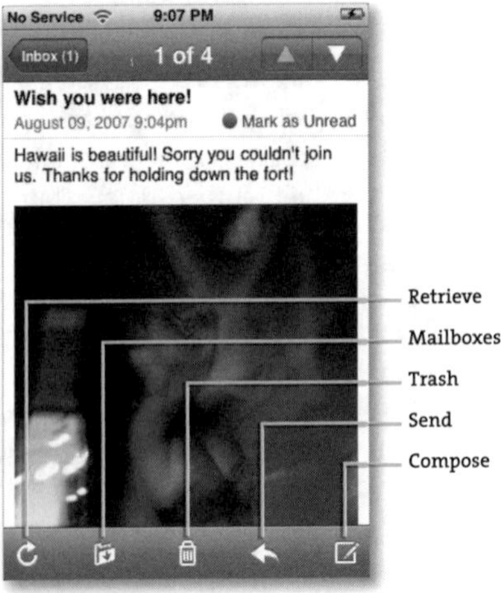

Figure 4.7 Message with attached photo, with Mail tool icons at the bottom of the screen.

URLs, email addresses, and phone numbers contained within messages will appear as blue, live links. Tap a URL, and Safari launches and takes you to that Web page. Tap an email address, and a new email message opens with that address in the To field. A tapped phone number causes a dialog box to appear. In it, you see the phone number and icons that offer to Cancel or Call.

tip **The iPhone is smart about URLs and phone numbers. Both http://www.examplesite.com and www.examplesite.com appear as live links, for example (but examplesite.com does not). And (555) 555-1212, 555-555-1212, 555.555.1212, 5555551212, and 5551212 are all live links and produce the Call dialog box when tapped.**

The tools below

The toolbar at the bottom of the screen contains five icons (see Figure 4.7):

- **Retrieve.** Tap this circular icon, and the iPhone will check for new messages for that account.
- **Mailboxes.** When you tap the Mailboxes icon, you're presented with a list of all the mailboxes associated with that account. Tap one of these mailboxes, and the message will be filed there. (Use this method to move a message out of the Trash.)
- **Trash.** Tap this icon, and the cute little trash can pops its top and sucks the message into it. Like I said, to move messages out of the Trash, just tap

the Trash mailbox in your account screen, tap a message, tap the Mailboxes icon, and then tap the mailbox where you'd like to put the message.

- **Send.** The left-arrow icon is your pathway to the Reply and Forward commands (**Figure 4.8**).

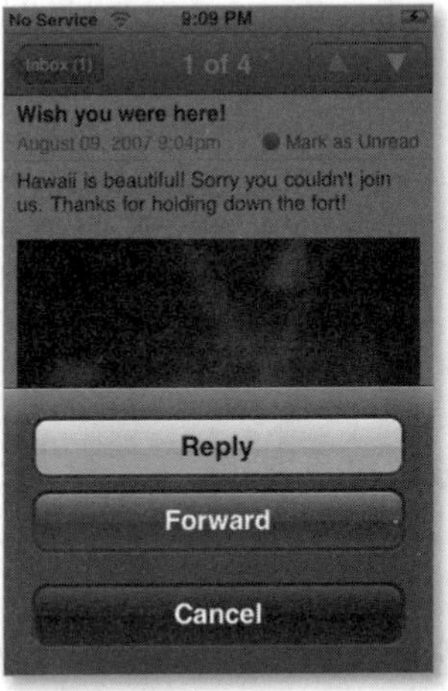

Figure 4.8
Reply button.

When you tap the Send icon and then the Reply button that appears, a new message appears, with the Subject heading *Re: Original Message Subject,* in which where *Original Message Subject* is ... well, you know. The message is addressed to the sender of the original message, and the Insertion point awaits at the top of the message body. The original text is quoted below. The message is mailed from the account you're working in.

Tap Forward, and you're responsible for filling in the To field in the resulting message. You can type it in yourself with the keyboard that appears or tap the plus (+) icon to add a recipient from your iPhone's list of contacts. This message bears *Fwd:* at the beginning of the Subject heading, followed by the original heading. The original message's From and To information appear at the top of the message as quoted text followed by the original message.

- **Compose.** Last is our old friend the Compose icon. Tap it, and a New Message screen appears, ready for your input.

Creating and sending email

If it truly is better to give than receive, the following instructions for composing and delivering mail from your iPhone should enrich your life significantly. With regard to email, the iPhone can give nearly as good as it gets. Here's how to go about it.

As I mention earlier in the chapter, you can create new email messages by tapping the Compose icon that appears on every account and mailbox screen. You'll even find the Compose icon available when you've selected Trash. To create a message, follow these steps:

1. Tap the Compose icon.

   Remember, the message will be sent from the account you're currently working in. If you've selected your CompanyX.com account, that account will send the message.

2. In the New Message screen that appears, type the recipient's email address or in the To field, tap the plus icon.

   When you place the cursor in the To or Cc field, notice that the iPhone's keyboard adds @ and period (.) characters where the spacebar usually resides. (The spacebar is still there; it's just smaller.) This makes it easy to type addresses, without having to switch to the numbers and symbols keyboard.

   When you start typing a name, the iPhone will suggest recipients based on entries in your list of contacts (**Figure 4.9**). If the recipient you want appears in the list below the To field, tap that name to add it to the field.

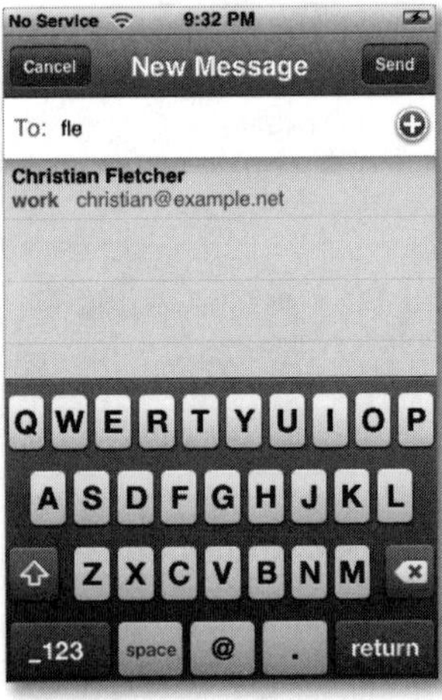

Figure 4.9 Begin typing to find a contact.

When you tap the plus icon, your list of contacts appears. Navigate through your contacts and tap the one you want to add to the To field. Some contacts will have multiple email addresses; tap the one you'd like to use. To add more names to the To field, type them or tap the plus icon to add them.

To delete a recipient, tap it and then tap the Delete key on the iPhone's keyboard.

3. If you'd like to Cc someone, tap in that field and then use any of the techniques in step 2 for adding a recipient.

4. Tap the Subject field, and enter a subject for your message with the iPhone's keyboard.

   That subject will replace the words *New Message* at the top of the screen.

tip **Use something original. Spam is a common problem in email, and some spam utilities will filter out messages with Subject headings such as "Hi!" or "Hello!" or the ever-popular "Replica Watches!".**

5. Tap in the message body (or, if the insertion point is in the Subject field, tap Return on the iPhone's keyboard to move to the message body), and type your message.

6. Tap Send to send the message or Cancel to save or delete your message.

   The Send icon, in the top-right corner, is easy enough to understand. Tap that icon, and the message is sent from the current account.

Cancel is a little more confusing. If you've typed anywhere in the New Message screen's Subject field or message body (even if you subsequently deleted everything you typed), a sheet will roll up when you tap Cancel, displaying Save, Don't Save, and Cancel icons. Tap Save to store the message in the account's Drafts mailbox. (If no such mailbox exists, the iPhone will create one.) When you tap Don't Save, the message is deleted. When you tap Cancel, the iPhone assumes that you made a mistake when you tapped Cancel the first time and it removes this sheet.

If the iPhone can't send a message—when you don't have access to a Wi-Fi network or AT&T's EDGE network, for example—it will create an Outbox for the account from which you're trying to send the message. When you next use Mail and are able to send the message, the iPhone will make the connection and send any messages in the Outbox, at which point the Outbox will disappear.

Using Calendar

Much like the iPhone's iPod area and Photos application, the Calendar application depends on the goodwill of your computer. Much of its heavy lifting is done with the aid of your Mac's or PC's calendar application and iTunes. Before examining just how heavy a load this is, I'll take a look at what you can do with Calendar without a computer's assistance.

Viewing events

Calendar is capable of displaying events in three views: List, Day, and Month. They're laid out like so.

Month

Tap Calendar, and by default, you'll see this month's calendar, with today's date highlighted in blue. Other days maintain a gray, businesslike appearance. Tap another day, and it adopts the blue box, while the present day gains a deeper gray hue. To return to the current day, either tap it (if you're viewing the current month) or tap the Today button in the top-left corner of the screen. To move to the next or previous month, tap the Previous or Next arrow, respectively, next to the month heading. To scan ahead more quickly, tap and hold one of these arrows.

tip

If you must know, the calendar will move as far back as January 1933 and ahead to December 2068. If you pick up this book in 2069 and are disappointed that the iPhone of 2007 won't allow you to create new events in your era of flying cars, consider upgrading. You've more than gotten your money's worth.

Any days on the calendar that have events appended to them bear a small black dot below the date. Tap a day with a dot, and the events for that day appear in a list below the calendar, each preceded by its start time (**Figure 4.10**). Tap an event in this list, and you're taken to the Event screen, which details the name and location of the event, its date, its start and end times, and any notes you've added to the event.

Figure 4.10 Month view with an event.

To edit or delete the event, tap the Edit icon in the top-right corner of the screen. Within the Edit screen, tap one of the fields to change its information. (I discuss these fields in "Creating events" later in this chapter.) To delete an event, tap the red Delete Event button at the bottom of the screen; then tap the Delete Event confirmation icon that appears.

Day

Tap the Day view button, and as you'd expect, you see the day laid out in a list, separated by hours. The day of the week and its date appear near the top of the screen. To move to the previous or next day, tap the Previous or Next arrow, respectively. To scan back or forward more quickly, tap and hold the appropriate arrow.

Events appear as blue-gray bars in the times they occupy, and are labeled with the name of the appointment and its location (**Figure 4.11**). Just as with events in Month view, tap them to reveal their details; to edit them, tap the Edit button.

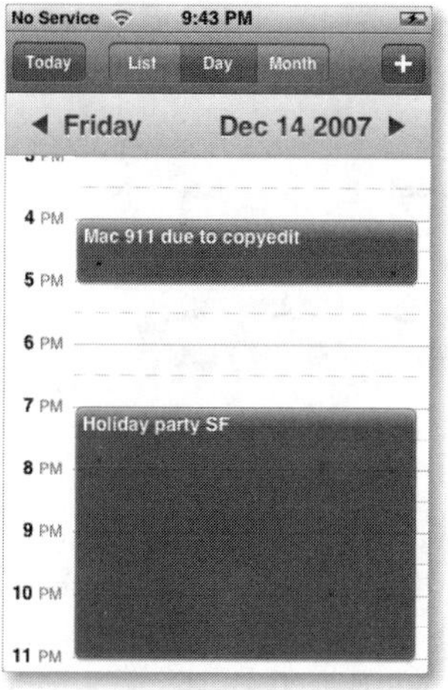

Figure 4.11 Day view with two events.

List

List view shows a list of all the events on your calendar , separated by gray bars. Each gray bar bears the day's abbreviated name (Fri or Mon, for example) and the month, date, and year of the event. The event's title appears just below, preceded by its start time. Once again, tap an event to view its details. Tap Edit to edit the event or delete it via the Delete Event button (**Figure 4.12**).

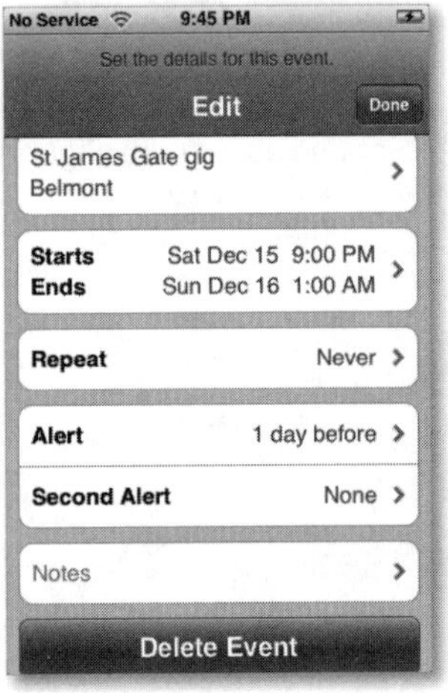

Figure 4.12 Editing an event.

Creating events

Creating events on the iPhone is simple. Just tap the plus icon in the top-right corner of the screen to produce the Add Event screen, where you'll find fields for Title & Location, Start & End, Repeat, Alert, and Notes. In more detail:

- **Title & Location.** The title of the event will appear when you select the event's date in Month view. Both an event's title and location appear in the Day-view list. And in List view, you see just the event's title. As with any other field on the iPhone, just type the entries and tap Save when you're done.

- **Start & End.** The title is explanation enough. Just tap the Starts field, and enter a date and time by using the spinning wheels at the bottom of the screen (**Figure 4.13**). Ditto with the Ends field. If the event lasts all day, tap the All-Day On/Off switch.

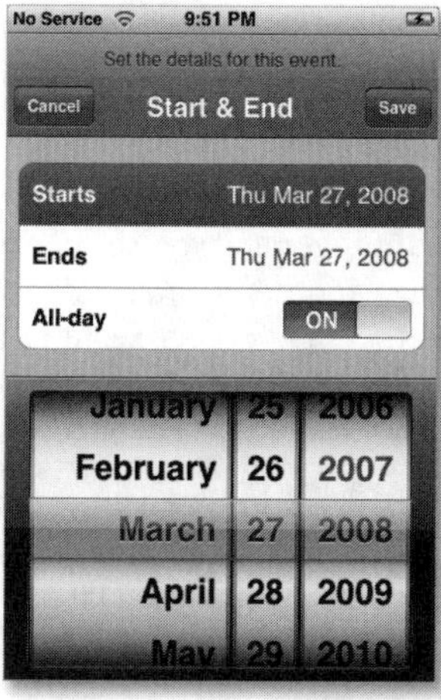

Figure 4.13 Set the duration of your event.

tip Unlike most calendar applications you're familiar with, this one lets you create events that span multiple days. Just dial in a different day when you tap Ends. The catch is that multiday events appear only in Day view. In Month and List views, they appear just on the day the event begins.

- **Repeat.** You can create an event that occurs every day, week, two weeks, month, or year. This method is a convenient way to remind yourself of your kid's weekly piano lesson or your own wedding anniversary.
- **Alert.** A fat lot of good an electronic calendar does you if you're not paying attention to the date or time. Tap Alert and direct the iPhone to sound an alert at a specific interval before the event's start time: 5, 15, or 30 minutes; 1 or 2 hours; 1 or 2 days before; or on the date of the event.

  You can create two alerts per event—useful when you want to remind yourself of events for the day and need another mental nudge a few minutes before the event occurs. Regrettably, you can't change the alert sound; you can only turn it on or off in the Sound Settings screen.
- **Notes.** Feel free to type a bit of text to remind you exactly why you're allowing Bob Whosis to dominate your Thursday afternoon.

Syncing events

Your computer and your iPhone have a nice sharing relationship with regard to events. When you create an event on one device, it's copied to the other, complete with title, location, start and end times, alerts (likely called *alarms* in your computer's calendar program), and notes.

Unfortunately, the iPhone holds just a single calendar. Although you can ask iTunes to sync multiple calendars to the iPhone (if you're using a Mac, your iCal Work and Home calendars, for example), all events go into that one iPhone calendar. And when the iPhone syncs events back to your computer, you have to tell iTunes which calendar to put them in.

tip

On my Mac, I've created a special iCal iPhone calendar where I place events I want to carry with me. Similarly, I tell iTunes to sync events I create on my iPhone to this iPhone calendar.

Deleting events

Quite frankly, deleting events by using the iPhone's interface is a pain in the neck. As I mention earlier in the chapter, you tap an event, tap the Edit button in the Event screen, tap the red Delete Event button at the bottom of the screen, and then tap Delete Event again. This is a very inefficient way to delete events, particularly lots of events that have expired. You're better off letting iTunes lend a hand.

To do so, plug your iPhone into the computer and then select it in iTunes' Source list. Click the Info tab, and configure the Calendars delete option to read *Do Not Sync Events Older Than X Days,* where *X* is the number of past days you're willing to keep expired events on your iPhone. When you next sync your iPhone, events that occurred more than *X* days before the current date will be removed from the phone (**Figure 4.14**).

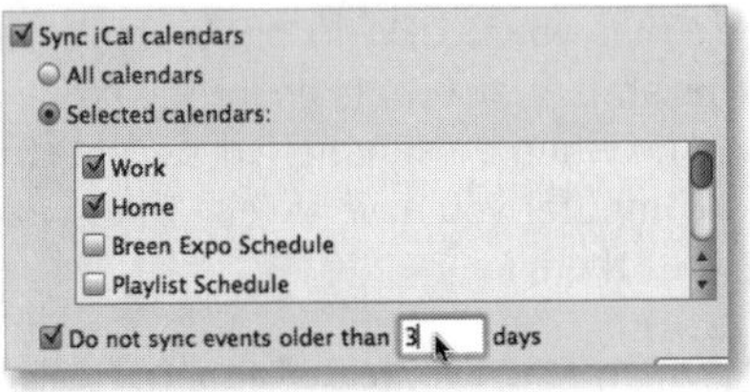

Figure 4.14 It's easier to delete lots of events through iTunes.

If you'd like to delete multiple future events , delete them from your computer's calendar. When you sync your iPhone, the events will disappear from the iPhone's calendar as well.

5

Safari

Got some free time? Great! Pick up your old cell phone, and look for the application it uses to browse the Web. No, really—if the phone was made in the past couple of years, it probably has such an application. Did you find it? Try over there. Ah, there you go.

What? Oh, that's right—you have to be signed up for a data plan to use it. Don't bother, really. By the time you do, and finally figure out how to get onto the Web, you'll likely see a very skimpy version of that very Web. That's the way phones—even smart ones—have done things in the past.

With the iPhone, that's no longer the case. Your phone has a real live Web browser, very much like the one on your computer. In this chapter, I show you how to use it to best advantage. Now let's go surfing!

Importing Bookmarks

I know you're eager to start surfing the Web with Safari, but you'll find the experience far more pleasant if you first sync your Safari (Mac) or Safari or Internet Explorer (Windows) bookmarks to your iPhone. This is easy to do:

1. Jack your iPhone into your computer's USB 2.0 port, launch iTunes (if it doesn't launch automatically), select the iPhone in iTunes' Source list, and click the Info tab.

2. In the Web Browser area of the window, on a Mac, enable the Sync Safari Bookmarks option (**Figure 5.1**); on a Windows PC, enable the Sync Bookmarks From option and choose either Safari or Internet Explorer from the pop-up menu.

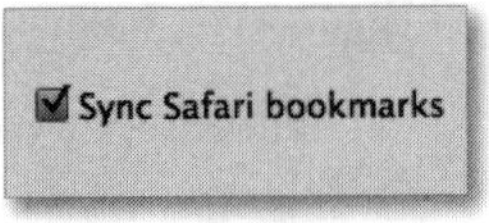

Figure 5.1 Syncing Safari within the Mac version of iTunes.

If you use a Web browser other than Safari or Internet Explorer, your browser undoubtedly has an option for exporting its bookmarks. (In Mozilla Firefox, for example, choose Bookmarks > Organize Bookmarks.) In the window that appears, choose File > Export; choose a location for saving your bookmarks; and click Save.

3. Now open Safari and choose File > Import Bookmarks, or fire up Internet Explorer and choose File > Import and Export.

4. Navigate to the bookmarks file you saved.

   Your bookmarks are now in a browser that's compatible with the iPhone. When you next sync your iPhone, those bookmarks will be available to the iPhone's copy of Safari.

Surfin' Safari

When you first tap the Safari icon at the bottom of the iPhone's Home screen, you may be surprised to see a full (though tiny) representation of a Web page appear before your eyes. Safari on the iPhone is nearly the real deal. (In "Safari's limits" later in the chapter, I talk about how that isn't quite the case.)

At first glance, though, it's the real *small* deal. The pages Safari displays on the iPhone are Lilliputian at first, but you have ways to make these pages legible:

- **Turn the iPhone on its side.** Yes, Safari is one of those iPhone applications that works in both portrait and landscape orientations. It displays the entire width of a Web page in either view, so when you switch to landscape orientation, you see more detail as the page enlarges to fill the iPhone's screen (**Figure 5.2** on the next page).

Figure 5.2 A Web page in landscape orientation, showing Safari's tool icons.

- **Stretch open the page.** You can enlarge the page by using the stretch gesture. When the page is enlarged, tap and drag to reposition it.
- **Double-tap a column.** Most Web pages include columns of text and graphics. To zoom in on a single column, double-tap it. That column will expand to fill the iPhone's screen. To shrink the page to its original size, double-tap the screen again.
- **Double-tap part of the page.** If a Web page lacks columns, you can still zoom in by double-tapping the page.

Browsing the Web

Like any good browser, Safari provides numerous ways to get around the Web. Let me count the ways.

Getting addressed

Like your computer's Web browser, Safari has an Address field at the top of its main window. To travel to a Web site, tap in this field. When you do, the iPhone's keyboard appears. If ever there were an argument for using Safari in landscape orientation, this feature is it, because the iPhone's keyboard is far less cramped this way (**Figure 5.3**). For just this reason, I wish that other applications supported landscape orientation.

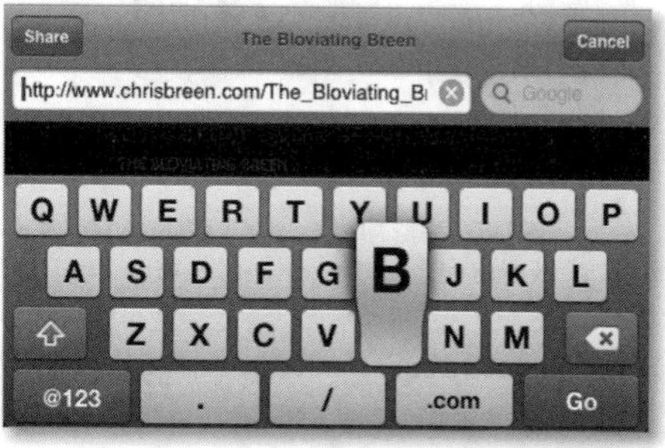

Figure 5.3
The landscape Safari keyboard.

Type the Web address you want to visit. The iPhone and its keyboard make this process as easy for you as possible. To begin with, you needn't type **http://www**. Safari understands that just about every Web address begins this way and doesn't require you to type the prefix. Just type **examplesite**; then tap the .com key at the bottom of the keyboard (even .com is unnecessary sometimes), and tap Go. In a short (Wi-Fi) or long (EDGE) time, the page you desire will appear.

Safari offers some other convenient shortcuts for entering addresses. If you've visited the site before, for example, it's likely to be in Safari's History list. If so, just begin typing the address, and it will appear below the Address field (**Figure 5.4**). Tap the address to go to that Web site.

Figure 5.4 The iPhone's History list can save typing.

tip **If the Address field is full when you tap it, you can erase its contents quickly by tapping the X icon that appears at the right edge of the field.**

note **If you're concerned that the contents of your iPhone's History list may give others pause, you can clear the list. See "Making Safari Settings" at the end of the chapter to learn how.**

If you need to type a more complex address—**example.com/pictures/vacation.html**, for example—the iPhone's default keyboard for Safari can help, because it includes both period (.) and slash (/) keys.

The keyboard offers a couple of other nifty features. If you find a Web site that you'd like to tell your friends about, tap the Share icon. An unaddressed email message appears, with the URL in the message body and with a Subject heading that includes the name of the site (**Figure 5.5**).

Figure 5.5 Share your favorite Web sites via email.

To leave the keyboard behind without doing anything, tap the Cancel button. If the page you're trying to visit is taking too long to load, or if you've changed your mind about visiting it, just tap the X that appears next to the Address field while the page is loading. Safari will stop loading the page. If you'd like to reload a page that's fully loaded, tap the Reload icon next to the Address field (the one that takes the place of the X when a page is completely loaded).

Search

You can also conduct Google or Yahoo searches from the keyboard. In portrait orientation, the Search field appears below the Address field; in landscape orientation, you'll find it to the right of the Address field. Tap the Search field, enter your query, and then tap Google or Yahoo (depending on which search engine you're using).

By default, the iPhone uses Google search. To switch to Yahoo, go to the Settings screen, and tap Safari. Tap Search Engine and then tap Yahoo.

Links

Links work in Safari just as they do in your computer's browser. Just tap a link to be taken to the associated Web page. Two things are worth noting:

- Safari is sometimes reluctant to use a link while it's still loading a Web page, which can be particularly infuriating when you're surfing the Web very slowly over EDGE. To speed things up, tap the X icon next to the Address field to stop the current page from loading; then tap the link to load its target immediately.
- When you hover your mouse pointer over a link in your computer's Web browser, you can view information about where that link will take you. The iPhone offers a similar, though hidden, capability. Just tap and hold a link, and the name of the link and its URL will appear in a gray bubble (**Figure 5.6**). This feature is useful when you suspect an innocuous-looking link may take you to a bad place.

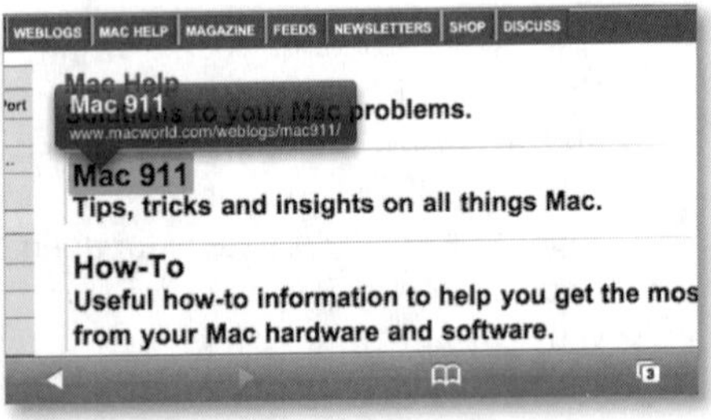

Figure 5.6 Preview the location of the link you're about to tap.

Back and forward

Just like your computer's Web browser, Safari has Back and Forward arrows for moving through sites you've visited.

Saved pages

In the bottom-right corner of the Safari screen, you'll see a small Pages icon. Tap it, and you'll see a small representation of the page you're currently viewing. Tap the New Page button in the bottom-left corner of the screen, and you can create a new empty Web page, saving the page you were just viewing in the process (**Figure 5.7**). This feature is the iPhone's equivalent of browser tabs.

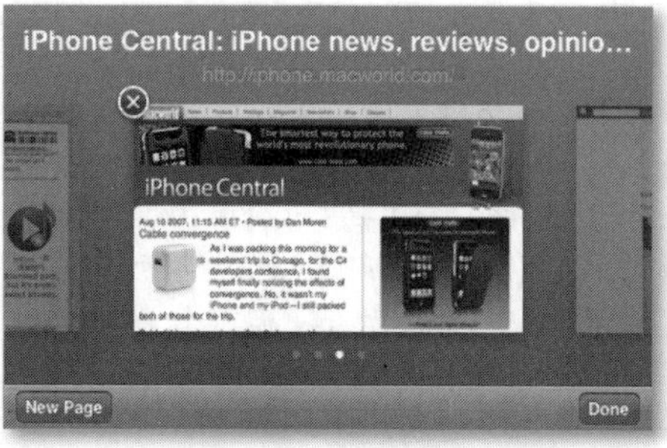

Figure 5.7
Safari lets you save up to eight pages.

You can repeat this process to save as many as eight pages, and the Pages icon will display the number of pages you've stored. To visit one of your saved pages, tap the Pages icon, and swipe your finger across the display to move back or forward through the saved pages. To view a page full-screen, tap its thumbnail or tap the Done button while its thumbnail is on view. To delete a page, tap the red X in the top-left corner of the page.

note **The contents of saved pages aren't cached to the iPhone—just their locations. So you won't be able to read them if your iPhone is offline (when you can't access a Wi-Fi or EDGE network, for example, or when your phone is in airplane mode).**

Navigating with bookmarks

You heeded my advice to import your computer browser's bookmarks, right? Great. Bookmarks are another fine way to get where you want to go.

Just tap the Bookmarks icon at the bottom of Safari's screen. The Bookmarks screen will appear, replete with your bookmarks organized as they were on your computer. By this, I mean that if you've organized your computer's bookmarks in folders, that's just how they'll appear on your iPhone. Bookmarks that you've placed in Safari's Bookmarks Bar are contained in their own folder, named (aptly enough) Bookmarks Bar.

Tap a folder to view the bookmarks it contains. To travel to a bookmark's target page, tap the bookmark.

More on bookmarks

Bookmarks are important-enough components of Safari that they deserve more than this so-far-brief mention. For example, how do you create bookmarks, organize and edit the ones you have, and delete those you no longer need? Like this.

Creating bookmarks

You've found a Web site you like while surfing with the iPhone. To bookmark the site, follow these steps:

1. Tap the plus (+) icon next to the Address field.
2. In the Add Bookmark screen that appears, check the name of the bookmark in the Title field.

   If the name is too long for your liking, edit it with the iPhone's standard text-editing techniques, or tap the X icon to erase the title and enter a title of your own.
3. Tap the Bookmarks entry, and choose a location for your bookmark.

   When you do this, a list that contains your bookmarks-folder hierarchy appears. Tap the folder where you'd like to file your bookmark. From now on, this folder is where you'll find that bookmark (**Figure 5.8**).

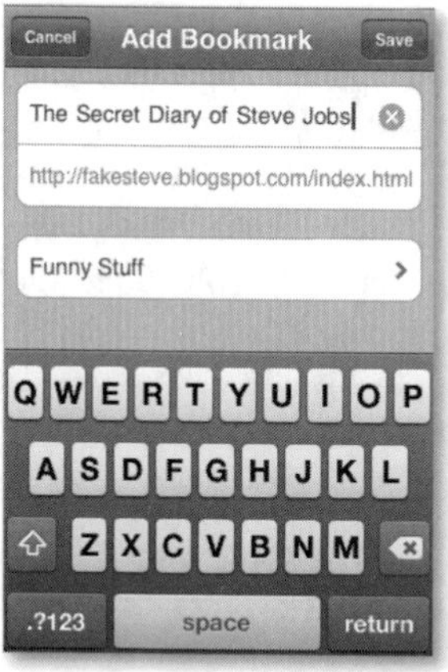

Figure 5.8 Creating a bookmark.

4. Tap Save to save the bookmark in this location, or tap Cancel to cancel the bookmarking operation.

Organizing and editing bookmarks

If you're as organized as I am (meaning not very), your bookmarks may be a bit of a mess. Although you're better off organizing the bookmarks on your computer and then syncing them to your iPhone, you can organize them on the phone as well. To do so, follow along:

1. Tap the Bookmarks icon.
2. In the resulting Bookmarks screen, tap the Edit button.
3. To delete an item, tap the red minus (–) icon that appears next to it.

The red minus icon appears next to all entries in the screen save History, Bookmarks Bar, and Bookmarks Menu—in short, all the items you've created but none of the items that the iPhone requires.

You'll also notice the three-line reposition icon to the right of these marked items, indicating that you can change their positions in the list by dragging the icons up or down. You can also rename your bookmark, change its URL, or file it in a different folder by tapping its name while in editing mode and then making those changes in the resulting Edit Bookmark screen.

Browser as application host

As you may have heard, third-party developers have no way to install their applications directly on the iPhone. Apple has shut that door. Web-based applications can work with the iPhone, however, and the key to making them work is Safari. Just launch Safari and travel to a site that acts as a host for iPhone Web applications, and your iPhone gets a whole lot more capable. (Provided, of course, that you're connected to the Web. Break that connection, and these Web-based applications no longer work.)

In the early days of the iPhone, some of these Web applications are just silly, but others are useful. Some allow you to use instant messaging, access more-advanced calculators and converters, play games, browse online dictionaries, or track flights. Pretty much anything that a person can do in a Web browser is finding its way to the iPhone.

Because books go out of date, rather than recommend specific Web applications that may be bested the day after this book sees print, I'll refer you to the iPhone Application List (**Figure 5.9** on the next page; http://iphoneapplicationlist.com), which keeps a constantly updated list of iPhone applications.

Figure 5.9 iPhone Application List keeps a current list of iPhone Web applications.

Safari and RSS

Safari supports RSS (Really Simple Syndication), the standard for distributing Web headlines. To view collections of these headlines (called *feeds*) on your iPhone, all you need to do is locate a page's RSS link and tap it. The page that appears bears a blue bar at the top, along with the name of the site connected to the feed—iPhone Central, for example (**Figure 5.10**). The site's headlines appear below the blue bar. Just tap a headline to read the full story.

Figure 5.10 A Safari RSS feed.

tip **RSS URLs are clumsy to enter yourself; they're long and rarely contain real words. For this reason, bookmarking those that you intend to revisit is a good idea.**

Safari's limits

Earlier in this chapter, I hinted that although the iPhone's version of Safari is about as full featured as you're likely to find on a mobile phone, it doesn't have all the capabilities of your computer's browser. The following sections discuss its limitations.

No Flash or Java support

Many modern Web sites greet you with luxurious animations, flickering icons, and animated menus created with Adobe's Web animation-design tool, Flash. The iPhone doesn't support Flash, and because it doesn't, you may see nothing at all on such a site's home page. Ideally, the designer took into account the fact that not everyone likes (or, in the case of the iPhone, can use) Flash and inserted a Skip Animation link that takes you to a Flashless version of the site.

Similarly, many of the movies you find on the Web are Flash-based. If, while traipsing through a Web site, you see a small blue box with a question mark inside it, you're looking at the placeholder for a Flash movie. Tapping that icon will do you no good whatsoever.

The good news is that the iPhone will play a lot of QuickTime content (though not all). As the iPhone increases in popularity, Web sites likely will increase their use of QuickTime.

No autofill

You're probably accustomed to your computer's browser automatically filling in information such as your user name, address, and phone number when you visit certain sites. The iPhone won't do this—and for good reason. If you lose your phone, do you really want the person who finds it to log into your Amazon and eBay accounts? I thought not.

No opening links in new pages

I mention earlier in the chapter that although you can open new Safari pages, you have no command for tapping a link and opening that link in a new page, much as you would in a browser that supports tabs.

No copy, paste, or downloading features

A global issue with the iPhone is that you can't copy or paste anything. And the iPhone version of Safari doesn't support downloading (because, in most cases, what good would it do you?). While I can get by without downloading, I miss copy and paste in a Web browser, particularly when I want to snag and save an image that I like or copy a bit of text from a page and drop it into Notes.

No Find

Web pages can be packed with information, and the iPhone's screen is a pretty small place to view that much content. I'd love to be able to pull up a Search field and key in a word or phrase I seek. I can't.

No Voice over IP

OK, this issue isn't technically a Safari issue, but it's a broad-enough Internet issue that it deserves to be noted here. Voice over IP (VoIP) services such as Skype aren't supported by the iPhone. Why? AT&T prefers that you make *real* calls with the iPhone—you know, *real* calls that count against your minutes.

Making Safari Settings

Like other iPhone applications, Safari has its own collection of settings. As you might guess, you find them by tapping Settings in the iPhone's Home screen and then tapping Safari in the Settings screen (**Figure 5.11**).

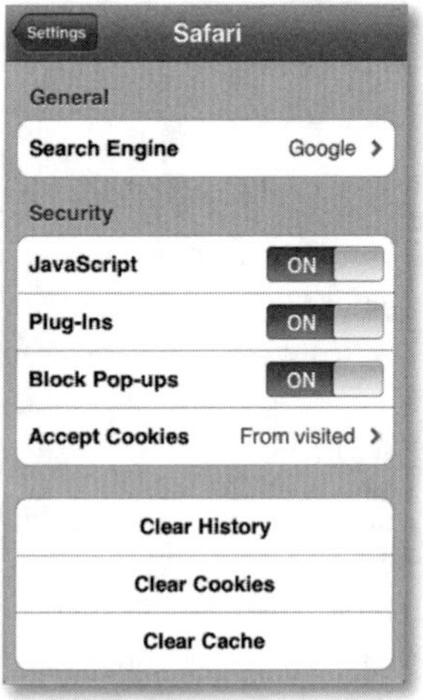

Figure 5.11
Safari settings.

These settings include the following:

- **Search Engine.** The iPhone can use either Google or Yahoo for its Web searches. Choose the one you want here.
- **JavaScript.** JavaScript is a scripting language that helps make Web sites more interactive. By default, Safari allows JavaScript to work. If you care to

disable JavaScript for some reason, you do it with this On/Off switch.

- **Plug-Ins.** The iPhone supports some plug-ins that allow it to display or play certain Web content—QuickTime movies and audio, for example. You can turn off these plug-ins by toggling this switch to Off.
- **Block Pop-Ups.** I make a lot of my living by writing for advertising-based Web sites, but I've yet to see a pop-up window that did more than annoy me. If you're haunted by pop-up ads, leave this option On.
- **Accept Cookies.** Many Web sites leave little markers called *cookies* stored in your Web browser. Cookies can store information such as when you visited the site and which pages you saw there. Sometimes, they also store information such as your user name and password for that site.

  The Accept Cookies setting gives you a measure of control over these cookies:

  - You can choose never to accept them (which some people consider to be more secure and private, but which forces you to reenter passwords and user names each time you return).
  - You can opt to accept just those cookies sent by each site you visit. (Some sites plant cookies from their advertisers, and this option prevents that behavior.)
  - You can choose always to accept cookies, which means that your iPhone is now a cookie-

gathering machine. The default setting is From Visited, which I think nicely balances privacy and convenience.

- **Clear History, Clear Cookies, and Clear Cache.** The final three buttons in the Safari Settings screen allow you to wipe your tracks:
  - Earlier, I told you that when you started typing a URL in Safari's Address field, the iPhone makes suggestions based on past searches. To stop this behavior, tap Clear History.
  - If you're concerned that the iPhone's stored cookies reveal more about your browsing habits than you're comfortable with, tap Clear Cookies.
  - Safari's cache stores some of the contents of pages you visit so that they open faster when you revisit. If new content isn't showing up, and you believe that it should, tapping Clear Cache will help by forcing Safari to reload entire pages that were previously cached.

When you tap any of these buttons, you're asked to confirm that you really want to perform the action.

6

iPhone as iPod

Whether you view your iPhone as a phone that just happens to play music and videos or as a really cool iPod that you can use to make calls, the fact remains that the iPhone's media capabilities are among its greatest. You can use your iPhone to listen to the best of your music collection; check out the latest podcasts, and watch your favorite TV shows, movies, and music videos. This chapter will show you how to do all that, as well as offer pointers for configuring iTunes to make the most of your iPhone's iPod functions.

Getting the Goods

"Aaack!" I hear you scream. "I've never used iTunes or owned an iPod. I have no idea how to get music into iTunes, much less put it on my iPhone. What do I do?"

Relax. I'm not going to tell you how to put your music and movies on the iPhone until you know how to assemble a music and movie library.

I'll start with music. You have three ways to get tunes into iTunes:

- Recording (or *ripping,* in today's terminology) an audio CD
- Importing music that doesn't come directly from a CD (such as an audio track you downloaded or created in an audio application on your computer)
- Purchasing music from an online emporium such as Apple's iTunes Store

The following sections tell you how to use the first two methods. Should you wish to explore the ins and outs of the iTunes Store, may I suggest that my *The iPod and iTunes Pocket Guide* is a fine addition to any library?

note **The procedures for adding movies and videos are similar, except that iTunes offers no option for ripping commercial DVDs. You can do that, but the procedure is more complicated than ripping an audio CD.**

Rip a CD

Apple intended the process of converting audio-CD music to computer data to be painless, and it is. Here's how to go about it:

1. Launch iTunes.
2. Insert an audio CD into your computer's CD or DVD drive.

   By default, iTunes tries to identify the CD you've inserted. It logs on to the Web to download the CD's track information—a very handy feature for those who find typing such minutia to be tedious.

   The CD appears in iTunes' Source list under the Devices heading, and the track info appears in the Song list to the right (**Figure 6.1**).

Figure 6.1 A selected CD and its tracks.

   Then iTunes displays a dialog box, asking whether you'd like to import the tracks from the CD into your iTunes Library.

3. Click Yes, and iTunes imports the songs; click No, and it doesn't.

note **You can change this behavior in iTunes' Preferences window. In the Importing tab of the Advanced pane, you find an On CD Insert pop-up menu. Make a choice from that menu to direct iTunes to show the CD, begin playing it, ask to import it (the default), import it without asking, or import it and then eject it.**

Figure 6.2 Import CD button.

4. If you decided earlier not to import the audio but want to do so now, simply select the CD in the Source list and click the Import CD button in the bottom-right corner of the iTunes window (**Figure 6.2**).

note **To import only certain songs, uncheck the boxes next to the titles of songs you don't want to import; then click the Import CD icon.**

iTunes begins encoding the files via the method chosen in the Importing tab of the Advanced pane of the iTunes Preferences window (**Figure 6.3**). By default, iTunes imports songs in "high quality" AAC format at 128 Kbps. (For more on encoding methods, see the sidebar "Import Business: File Formats and Bit Rates.")

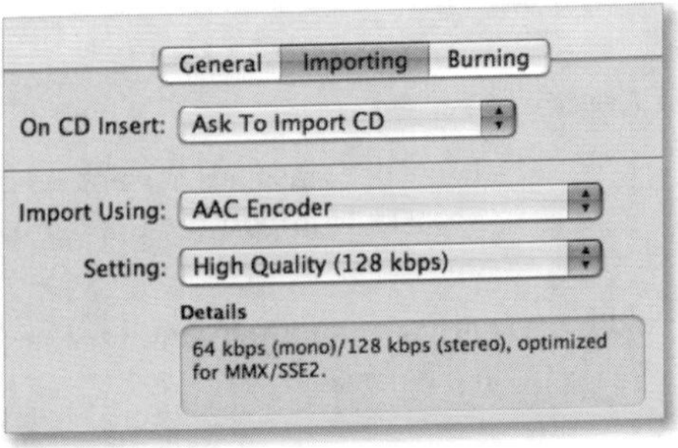

Figure 6.3 iTunes' Importing tab.

5. Click the Music entry in the Source list.

You'll find the songs you just imported somewhere in the list.

6. To listen to a song, click its name in the list and then click the Play icon or press the spacebar.

Import Business: File Formats and Bit Rates

MP3, MPEG-4, AAC, AIFF, WAV ... is the computer industry incapable of speaking plain English?

It may seem so, given the plethora of acronyms floating through modern-day Technotopia. But the lingo and the basics behind it aren't terribly difficult to understand.

MP3, AAC, AIFF, and WAV are audio file formats. The compression methods used to create MP3 and AAC files are termed *lossy* because their encoders remove information from the source sound file to create these smaller files. Fortunately, these encoders are designed to remove the information you're least likely to miss—audio frequencies that humans can't hear easily, for example.

AIFF and WAV files are *uncompressed,* which means that they contain all the data in the source audio file. When a Macintosh pulls audio from an audio CD, it does so in AIFF format, which is the native uncompressed audio format used by Apple's QuickTime technology. WAV is an AIFF variant used extensively with the Windows operating system.

iTunes supports one other compression format: Apple Lossless. This format is termed a *lossless* encoder because it shrinks files by removing redundant data without discarding any portion of the

(continued on next page)

Import Business: File Formats and Bit Rates (continued)

audio spectrum. This scheme yields sound files with all the audio quality of the source files at around half their size.

iTunes and the iPhone also support the H.264 and MPEG-4 video formats. These, too, are compressed formats that allow you to fit a great big movie on a tiny iPhone.

Now that you're familiar with these file formats, I'll touch on *resolution* as it applies to audio and video files.

You probably know that the more pixels per inch a digital photograph has, the crisper the image (and the larger the file). Resolution applies to audio as well. But audio defines resolution by the number of kilobits per second (kbps) contained in an audio file. *With files encoded similarly,* the higher the kilobit rate, the better-sounding the file (and the larger the file).

I emphasize *with files encoded similarly* because the quality of the file depends a great deal on the encoder used to compress it. Many people claim that if you encode a file at 128 kbps in both the MP3 and AAC formats, the AAC file will sound better.

The Import Using pop-up menu lets you choose to import files in AAC, AIFF, Apple Lossless, MP3, or WAV format. The Setting pop-up menu is where you choose the resolution of the AAC and MP3 files encoded by iTunes by choosing Custom from the menu. iTunes' default setting is High Quality (128 kbps). To change this setting, choose Higher Quality (256 kbps) or Custom from the Setting pop-up menu. (Spoken Podcast is another option when you choose the AAC Encoder, but it produces quality that's good only for spoken word). If you choose Custom, the AAC Encoder dialog box will appear. Choose a different setting—in a range from 16 kbps to 320 kbps—from the Stereo Bit Rate pop-up

Import Business: File Formats and Bit Rates (continued)

menu (**Figure 6.4**). Files encoded at a high bit rate sound better than those encoded at a low bit rate (such as 96 kbps). But files encoded at higher bit rates also take up more space on your hard drive and iPhone.

Figure 6.4
The Stereo Bit Rate pop-up menu.

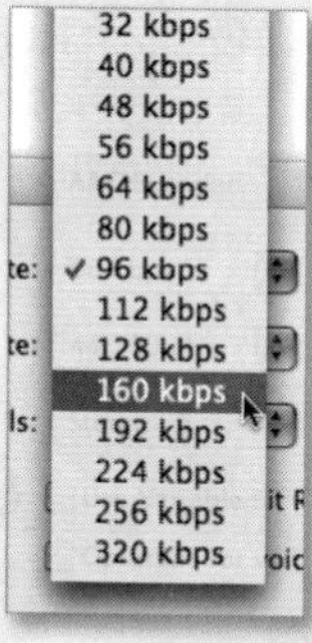

The preset options for MP3 importing include Good Quality (128 kbps), High Quality (160 kbps), and Higher Quality (192 kbps). If you don't care to use one of these settings, choose Custom from this same pop-up menu. In the MP3 Encoder dialog box that appears, you have the option to choose a bit rate ranging from 8 kbps to 320 kbps.

Resolution is important for video as well. Fortunately (because an explanation beyond this gross simplification is beyond the scope of this slim volume), iTunes doesn't require or even allow you to muck with encoding video in any way, shape, or form. Movies are either encoded in such a way that they play on your iPhone, or they aren't.

Move music into iTunes

Ripping CDs isn't the only way to put music files on your computer. Suppose that you've downloaded some audio files from the Web and want to put them in iTunes. You have three ways to do that:

- In iTunes, choose File > Add to Library.

  When you choose this command, the Add To Library dialog box appears. Navigate to the file, folder, or volume you want to add to iTunes, and click Choose (**Figure 6.5**). iTunes determines which files it thinks it can play and adds them to the library.

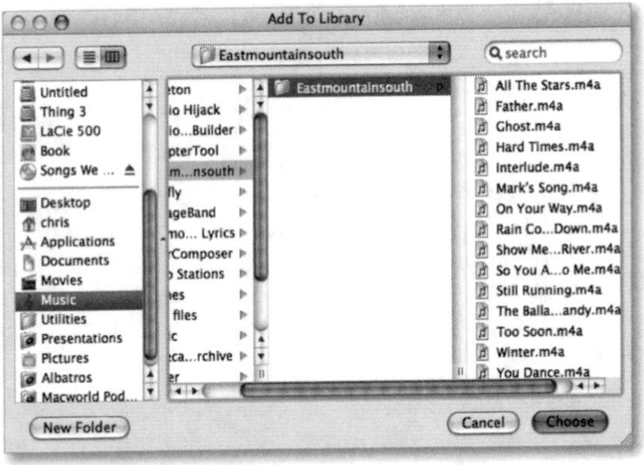

Figure 6.5 Navigate to tracks you want to add to iTunes via the Add To Library dialog box.

- Drag files, folders, or entire volumes to the iTunes icon in Mac OS X's Dock, the iTunes icon in Windows' Start menu (if you've pinned iTunes to this menu), or the iTunes icon in either operating system (at which point iTunes launches and adds the dragged files to its library).

- Drag files, folders, or entire volumes into iTunes' main window or the Library entry in the Source list.

  In the Mac versions of iTunes, by default you'll find songs in the iTunes Music folder within the iTunes folder inside the Music folder inside your Mac OS X user folder. The path to my iTunes music files, for example, would be chris/Music/iTunes/iTunes Music.

  Windows users will find their iTunes Music folder by following this path: ***yourusername***/My Music/iTunes/iTunes Music.

  You can use the same methods to add compatible videos and movies to your iTunes Library. (For more on what makes those videos compatible, see "Working with supported video formats" later in the chapter.) Those videos will most likely appear in the Movies playlist in the Source list.

  I say "most likely" because there are a few exceptions: Videos specifically designated as music videos appear in the Music playlist, and videos designated as TV shows appear in the TV Shows playlist. See the sidebar "Tag, You're It" later in this chapter for information on how to apply those video designations.

Creating and Configuring a Playlist

Before you put any media files (music or video) on your iPhone, organize them in iTunes. Doing so will make it far easier to find the media you want, both on your computer and on your iPhone. The best way to organize that material is through the use of playlists.

A *playlist* is simply a set of tracks and/or videos that you believe should be grouped in a list. The organizing principle is completely up to you. You can organize songs by artist, by mood, by style, by song length ... heck, if you like, you can have iTunes automatically gather all your 1950s polka tunes with the letter *z* in their titles. Similarly, you can organize your videos by criteria including director, actor, and TV-series title. You can mix videos and music tracks within playlists as well, combining, say, music videos and music tracks by the same artist. As far as playlists are concerned, you're the boss.

The following sections look at ways to create playlists.

Standard playlists

Standard playlists are those that you make by hand, selecting each of the media files you want grouped. To create a standard playlist in iTunes, follow these steps:

1. Click the large plus (+) icon in the bottom-left corner of the iTunes window, or choose File > New Playlist (Command-N on the Mac, Ctrl-N in Windows).

2. In the highlighted field that appears next to that new playlist in the Source list, type a name for your new playlist (**Figure 6.6**).

Figure 6.6
Enter a name for your playlist.

3. Click an appropriate entry in the Source list—Music, Movies, TV Shows, or Podcasts—and select the tracks or videos you want to place in the playlist you created.

4. Drag the selected tracks or videos to the new playlist's icon.

5. Arrange the order of the tracks or videos in your new playlist.

   To do this, click the Number column in the main window, and drag tracks up and down in the list. When the iPhone is synchronized with iTunes, this order is how the songs will appear in the playlist on your iPhone.

   If the songs in your playlist come from the same album, and you want the songs in the playlist to appear in the same order in which they do on the original album, click the Album heading.

Playlist from selection

You can also create a new playlist from selected items by following these steps:

1. Command-click (Mac) or Ctrl-click (Windows) songs or videos to select the files you'd like to appear in the new playlist.
2. Choose File > New Playlist from Selection (Command-Shift-N on a Mac; the Windows version of iTunes has no keyboard shortcut).

   A new playlist containing the selected items will appear under the Playlists heading in the iTunes Source list. If all selected tracks are from the same album, the list will bear the name of the artist and album. If the tracks are from different albums by the same artist, the playlist will be named after the artist. If you've mixed tracks from different artists or combined music with videos, the new playlist will display the name *untitled playlist*.
3. To name (or rename) the playlist, type in the highlighted field.

Smart Playlists

Smart Playlists are slightly different beasts. They include tracks that meet certain conditions you've defined—for example, OutKast tracks encoded in AAC format that are shorter than 4 minutes. Here's how to work the magic of Smart Playlists:

1. In iTunes, choose File > New Smart Playlist (Command-Option-N on the Mac, Ctrl-Alt-N in Windows).

   You can also hold down the Option key on the Mac or the Shift key on a Windows PC and click the Gear icon that replaces the plus icon at the bottom of the iTunes window.

2. Choose your criteria.

   You'll spy a pop-up menu that allows you to select items by various criteria—including artist, composer, genre, podcast, bit rate, comment, date added, and last played—followed by a Contains field. To choose all songs by Elvis Presley and Elvis Costello, for example, you'd choose Artist from the pop-up menu and then enter **Elvis** in the Contains field.

   You can limit the selections that appear in the playlist by minutes, hours, megabytes, gigabytes, or number of songs. You may want the playlist to contain no more than 2 GB worth of songs and videos, for example.

   You'll also see a Live Updating option. When it's switched on, this option ensures that if you add any songs or videos to iTunes that meet the criteria you've set, those files will be added to the playlist. If you add a new Elvis Costello album to iTunes, for example, iTunes updates your Elvis Smart Playlist automatically.

3. Click OK.

A new playlist that contains your smart selections appears in iTunes' Source list.

You don't have to settle for a single criterion. By clicking the plus icon next to a criterion field, you can add other conditions (**Figure 6.7**). You could create a playlist containing only songs you've never listened to by punk artists whose names contain the letter *J*.

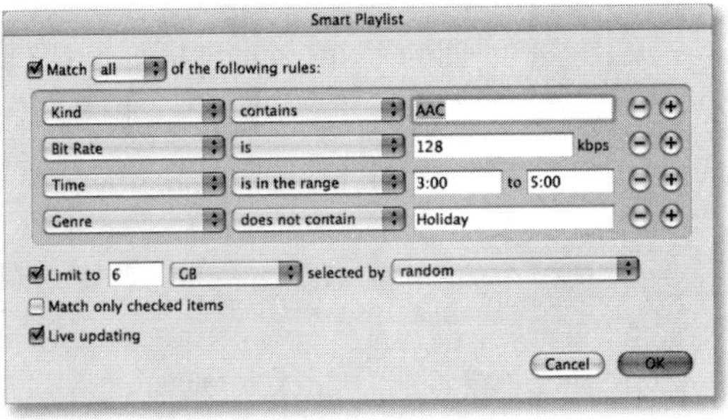

Figure 6.7
A Smart Playlist.

iTunes includes five Smart Playlists: 90's Music, My Top Rated, Recently Added, Recently Played, and Top 25 Most Played. These playlists have the Live Updating option enabled, which tells them to include new media files dynamically if they meet your search conditions. This feature lets your playlist reflect changes in your song ratings, addition of new tracks by a particular artist, and so on.

To see exactly what makes these playlists tick, Mac users can Option-click a Smart Playlist to open it. Windows users simply right-click a playlist to see this command.

Tag, You're It

So how does iTunes know about tracks, artists, albums, and genres? Through something called *ID3 tags*. ID3 tags are just little bits of data included in a song file that tell programs like iTunes something about the file—not just the track's name and the album it came from, but also the composer, the album track number, the year it was recorded, and whether it is part of a compilation.

These ID3 tags are the key to creating great Smart Playlists. To view this information, select a track and choose File > Get Info. Click the Info tab in the resulting window, and you'll see fields for all kinds of things. You may find occasions when it's helpful to change the information in these fields. If you have two versions of the same song—perhaps one is a studio recording and another a live recording—you could change the title of the latter to include *(Live)*.

A really useful field to edit is the Comments field. Here, you can enter anything you like and then use that entry to sort your music. If a particular track would be great to fall asleep to, for example, enter **sleepy** in the Comments field. Do likewise with similar tracks, and when you're ready to hit the hay, create a Smart Playlist that includes "Comment is sleepy." With this technique under your belt, you can create playlists that fit particular moods or situations, such as a playlist that gets you pumped up during a workout.

The Comments field can be useful for sorting movies as well. If you like a particular actor or director, enter his or her name in the Comments field—**Bogart** or **Huston,** for example.

Size matters

Unlike a full-size iPod—which, as I write this chapter, can hold up to 80 GB of stuff—the iPhone doesn't give you a lot of storage space to work with. The most expansive iPhone carries a fairly scant-by-iPod-standards 8 GB of storage capacity. For this reason, you have to be choosy about what you sync to your iPhone. I recommend these strategies:

- **Use compressed music.**

  Audio purists prefer their music in uncompressed form because it sounds better. But if you hope to pack your iPhone not only with music, but also a few TV episodes and a movie or two, you need to keep everything as slim as possible. You'll find that 128 Kbps AAC files don't sound half bad—particularly over the iPhone's headset, which offers sound that's of an acceptable quality but that's hardly designed for the audiophile. The protected music that the iTunes Store sells is 128 kbps. If you like its quality, you'll be fine ripping your CDs that way too.

- **Sync TV shows and movies first.**

  If you plan to watch TV shows and movies on your iPhone, first perform a sync with just those items selected in iTunes' Video pane. See how much room you have left; then create a Smart Playlist that will fit comfortably in the remaining space.

- **Delete files when you're done with them.**

  When a video concludes, the iPhone will offer to delete it (**Figure 6.8**). Take advantage of this opportunity; it's unlikely that you're going to watch that movie or TV show again before the next time you sync your iPhone.

Figure 6.8 The iPhone offers to delete videos after you've watched them.

When you do this with TV shows, remember to use the "unwatched" option when syncing your shows—choose Sync 3 Most Recent *Unwatched* Episodes of "Lost" rather than 3 Most Recent Episodes of "Lost," for example. The reason? When an episode concludes, the iPhone marks it as watched and tells iTunes so when it syncs. If you choose 3 Most Recent Episodes, iTunes will put those same three episodes back on your iPhone (unless a new episode has arrived since you last synced) when you next sync, even if you deleted them on the iPhone after watching them.

tip **Likewise with movies, be sure to uncheck any movie you've deleted from the iPhone. Even though you've deleted it, iTunes will sync it with the iPhone if it's checked in the Video tab.**

- **Use Play Count to keep material fresh.**

  If you have a large music collection, you'll likely want to rotate different music on and off your iPhone with each sync. To help you do that, create a Smart Playlist that uses Play Count as a condition. A Smart Playlist that reads *Play Count is o* with the Live Updating option enabled, for example, ensures that after you've played a track on your iPhone and sync it to iTunes, that track will be deleted from the Smart Playlist, and another will take its place.

- **Have reasonable expectations.**

  You can find a lot of great podcasts out there, but you have just so many listening hours in the day. When syncing podcasts, think about what you can really listen to in the period of time before you next sync your iPhone. Then choose a reasonable syncing option—just your three favorite podcasts and only the most recent unplayed episode, for example. This method is particularly important for video podcasts, which can consume a lot of storage space.

Using the iPod on iPhone

Now that you've filled your iPhone with great content, you'd probably like to know how to find and play it. Follow along as I walk through the iPod area of the iPhone.

Cover Flow view

Tap the orange iPod icon near the bottom-right corner of your iPhone's Home screen, wait for the Playlists screen to appear (which it does by default when you first tap iPod), and immediately turn the iPhone on its side. You're witnessing the iPhone's Cover Flow view, which lets you browse your music collection and podcasts by their album or program artwork (**Figure 6.9**). I don't care if you never choose to browse your music this way; Cover Flow is the iPod feature you'll choose first to impress your friends. They can't help but *oooh* in awe when you flick your finger across the screen and the artwork flips by.

Figure 6.9
Cover Flow view.

Should you want to navigate your music or podcasts in Cover Flow view, you can do so easily:

1. Turn the iPhone to landscape orientation (it doesn't matter whether this places the Home button on the right or left side, it works either way), and flick your finger across the display to move through your audio collection.

   Albums are sorted by the artist's first name, so Al Green appears near the beginning and The Weepies appears close to the end.

2. When you find an album you want to listen to, tap its cover.

   The artwork flips around and reveals the track list of the album's contents or, in the case of a podcast, the podcast episodes (**Figure 6.10** on the following page).

Figure 6.10 A track list in Cover Flow view.

As with other lists on the iPhone that may be longer than the screen, you're welcome to flick your finger up the display to move down through the list.

3. Tap the track you want to listen to.

   Playback begins from that track and plays to the end of the list in the order presented in the track list.

   To adjust volume in this view, use the Volume buttons on the side of the phone. To pause playback, tap the Pause symbol in the bottom-left corner of the screen or, if you're listening with the iPhone's headset, press the mic button once.

4. To move to another album, tap the album-art thumbnail in the top-right corner of the cover, double-tap an empty spot in the track list, or tap the *i* icon in the bottom-right corner of the screen.

   Any of these actions will flip the track list back to the artwork.

note

While you're listening to the contents of one album, you're free to view the contents of another. Just flick your finger across the screen to move through your collection. Go ahead and tap an album or podcast to see its contents. It won't play until you tap a track.

Play screen

Turn your iPhone so that it's in portrait orientation, and Cover Flow disappears; it works only in landscape orientation. What you're left with when you flip the iPhone to portrait orientation is the Play screen. This screen is what you'll use to perform several tasks, including navigating through an album, fast-forwarding, switching on shuffle or repeat play, and rating your tracks.

The Play screen has two main views, standard play and track list.

Standard play

The view you first see is straightforward. From the bottom of the screen to the top, you'll see a volume slider; play controls that include Previous/Rewind, Play/Pause, and Next/Fast Forward icons; album art; a Back icon; artist, track title, and album title information; and a Track List icon (**Figure 6.11** on the next page).

The volume slider operates like its real-world equivalent. Just drag the silver ball on the slider to the right to increase volume and to the left to turn the volume down. (You can use the iPhone's mechanical volume buttons to adjust volume as well.)

Figure 6.11 The music Play screen.

The Previous/Rewind icon earns its double name because of its two jobs. Tap it once, and you're taken to the beginning of the currently playing track or chapter of the currently playing podcast or audiobook. Tap it twice, and you move to the previous track or chapter. Tap and hold, and it causes the current playing track to rewind.

The Play/Pause icon toggles between these two functions.

The Next/Fast Forward icon works like Previous/Rewind: Tap once to move to the next track in the track list or chapter in an audiobook or podcast; press and hold to fast-forward through the currently playing track.

I'll skip album art for a second and move to the Back icon in the top-left corner of the screen. Tap this icon, and you'll move to the currently selected track view screen. If you've chosen to view your music by playlist, for example, you'll see your list of playlists. When you tap the Back icon and are taken to one of these screens, a Now Playing button appears in the top-right corner of the current screen. This icon appears whenever you're in the iPod area, making it easy to move to the Play screen.

Track list

In the top-right corner of the Play screen is the Track List icon. Tap this icon, and you get that album-cover flip effect again and a list of the current album's contents (**Figure 6.12**). (Naturally, if you have only a couple of tracks from that album stored on your iPhone, you'll see just those tracks.) Just as you can in Cover Flow view, you can tap an entry in the track list to listen to that track. Again, tracks play in order from where you tapped.

Figure 6.12
A track list in the music Play screen.

The Track List screen also includes a means for rating tracks. Just above the track list, you'll see five gray dots. To assign a star rating from 1 to 5, simply tap one of the dots. Tap the fourth dot, for example, and the first four dots turn to stars. You can also wipe your finger across the dots to add or remove stars. These ratings are transferred to iTunes when you next sync your iPhone. Tap the artwork image to flip the track list and return to the Play screen.

Additional controls

While you're in the Play screen, tap the artwork in the middle of the screen, and additional controls drop down from above (**Figure 6.13**). Starting from the left, you'll find a Repeat icon. Tap this icon once, and the contents of the currently playing album, audiobook, or podcast will repeat from beginning to end. Tap the Repeat icon twice, and just the currently playing selection will repeat.

Figure 6.13 Additional controls in the music Play screen.

A timeline with playhead comes next. To its left is the location of the playhead in minutes and seconds—1:40, for example. To its right is the track's remaining time. Drag the playhead with your finger to move to a different position in the currently playing track.

To the far right is the Shuffle icon. Tap this icon once so that it turns blue and the contents of the current album are shuffled; tap it again to turn shuffle off.

iPod content views

The iPhone's iPod area provides several ways to organize your media. Look across the bottom of the screen when you're in the iPod area (anywhere but in the Play screen), and you'll see five icons for doing just that: Playlists; Artists; Songs; Videos; and More, which leads you to even more options (**Figure 6.14**).

Figure 6.14 Category icons at the bottom of the iPod screens.

These icons are largely self-explanatory. When you tap Playlists, you'll see a list of all the playlists you've synced to your iPhone. Tap a playlist to move to a screen where all the tracks on the playlist appear in the order in which they were arranged in iTunes. If you tapped the Album heading when the playlist was displayed in iTunes, for example, the tracks will appear in that order; tap the Time heading, and the shortest tracks appear first on through to the longest tracks. Tap a track, and you're taken to the Play screen, where the track begins playing.

Whenever you choose a list of tracks in one of these views, Shuffle is the entry at the top of the list. Tap Shuffle, and the contents of that collection of tracks will play in random order.

The On-The-Go Playlist

Like the iPod, the iPhone lets you create an *On-The-Go playlist*, which you can create directly on the iPhone rather than syncing it from iTunes. You can add individual songs or clumps of songs to this special playlist. It works this way:

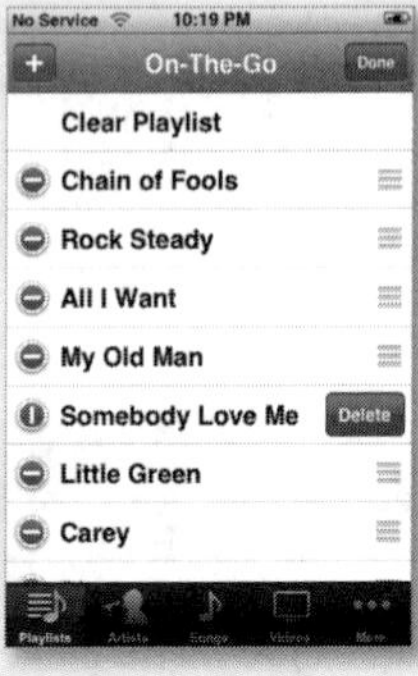

Figure 6.15 Editing the On-The-Go screen.

Tap On-The-Go, and a screen rises up from the bottom of the display, hinting that you've entered a special area of the iPhone (**Figure 6.15**). Tap one of the entries at the bottom of the screen: Playlists, Artists, Songs, Videos, or More (and then one of the selections available in the More screen). When you do, you'll see a screen that features the words *Add All Songs* followed by a list of all the songs that belong to that entry (all the songs on an album or by a particular artists, for example). If you tap Add All Songs, you do just that. To add individual songs, tap them. Continue tapping icons at the bottom of the screen or in the More screen until you've added all the tracks you care to; then tap Done at the top of the screen.

When you return to the Playlists screen and tap On-The-Go, you'll see a list of all the tracks you've added through your previous endeavors. To edit the contents of this playlist, tap the Edit icon in the top-right corner of the screen. In the Edit screen, you can tap the Clear Playlist entry to do just that; tap the minus (–) icon next to a track to produce the Delete icon, which allows you to remove that track from the playlist (but not from your iPhone), and the List icon, which you drag to change the position of the selected track in the playlist.

While you're in the Edit screen, you can also tap the plus icon to add more tracks to your On-The-Go playlist. Tap plus, and you're back to the view where you can add playlists, artists, songs, and so on.

Tap Artists. and you're presented with an alphabetical list of the artists represented on your iPhone. If your iPhone has tracks from more than one album by the selected artist, when you tap the artist's name, you'll be taken to an Albums screen that displays the titles of the artist's represented albums (along with thumbnails of their cover art). To view tracks from a particular album, tap its name. To view all songs by the artist, tap All Songs in this screen.

The Songs screen lists all the songs on your iPhone. Like any list that contains several dozen (or more) entries, this one displays a tiny alphabet along the right side of the screen. To navigate to a letter quickly, tap it (as best you can, as the letters are really small) or slide your finger along the alphabet listing to dash through the list.

note **If the first word of a list entry is *A* or *The*, the second word in the entry is used for sorting purposes. *The Beatles* is filed under *B*, for example, and *A Case of You* appears under *C*.**

The Video icon is your gateway to playing movies, TV shows, music videos, and music podcasts on your iPhone. I talk about playing videos later in this chapter.

When a Video Is Not a Video

In iTunes, you can create a playlist that contains both audio and video. iTunes will warn you that this isn't a good idea, but you can still do it. When you sync one of these playlists to your iPhone, one of two things will happen:

- If the video you added to the playlist is anything but a music video or a video podcast, it won't be synced to the iPhone. To sync TV shows and movies, you must choose them in the iPhone Preferences window's Video pane.
- If you've included a music video or a video podcast, it will be synced to your iPhone, but it will appear twice. One version will appear in the audio playlists—in the playlist you created that appears in the Playlists screen, under the artist's name in the Artists screen, and in the Songs screen. A video podcast also appears in the Podcasts screen. This version will play the video's audio track only. A frame from the video will be used for the artwork on the Play screen. The version that has actual moving video will appear in the iPhone's Video screen and will be filed under its type: Music Videos or Podcasts.

I know it sounds goofy, but there's a method to this apparent madness. This scheme allows you to enjoy the music in a music video (which presumably is at least part of the reason you have it) when you can't be bothered with video—when you're driving, for example. Same goes with a video podcast. You can listen to the content without having to gaze on a face best enjoyed on radio.

The iPhone has limited space, yet you have many more ways to organize your music—by albums, audiobooks, compilations, composers, and genres, for example. That's exactly what the More icon is for.

Tap it, and you'll see just those items I list, as well as a Podcasts entry. Tap these entries, and most of them behave pretty much as you'd expect—with a couple of variations.

The Albums screen, for example, lists albums in alphabetical order and displays a thumbnail of the cover art next to the name of each album.

The Compilations entry lists only those albums that iTunes denotes as compilations. These items are usually greatest-hits collections, soundtracks, or albums on which lots of artists appear—tribute albums or concert recordings, for example.

The Podcasts screen displays all the podcasts on your iPhone, along with their cover art. Tap a podcast title, and you're taken to a screen that lists all that podcast's episodes. Blue dots denote podcasts that you haven't listened to yet.

More Mucking

Unhappy that Apple chose to tuck the Albums entry in the More screen, yet left Artists easily accessible at all times? No worries. You can change what appears at the bottom of the iPod area. Simply tap More and then the Edit icon in the top-left corner of the screen. Doing so produces a Configure screen that swipes up from the bottom of the display. Here, you'll see all the iPod category entries listed. Find one you like, and drag it over a icon on the bottom of the screen that you want to replace. The new entry takes the place of the old one, and the old entry is listed in the More screen. When you're done, tap Done.

Videos

Like the latest full-size iPods, the iPhone plays videos. Some people would say that *unlike* these iPods, the iPhone makes videos actually watchable—bright and plenty big enough for personal viewing. Here are the ins and outs of iPhone video.

Working with supported video formats

Regrettably, you can't take just any video you pull from the Web or rip from a DVD and plunk it on your iPhone. Like the iPod, the iPhone has standards the video must meet before it's allowed to touch your iPhone.

Specifically, the videos must be in either MPEG-4 or H.264 format and must fit within these limits:

MPEG-4

Resolution: 640 by 480 pixels
Data rate: Up to 2.5 Mbps
Frame rate: 30 fps
Audio: Up to 48 kHz

H.264

Resolution: 640 by 480 pixels
Data rate: Up to 1.5 Mbps
Frame rate: 30 fps
Audio: Up to 48 kHz

You can also encode H.264 movies at a resolution of 320 by 240 at 30 fps. When you do so, the data rate is limited to 768 Kbps.

What? If you have experience encoding video, these numbers will make sense to you; if they have you confused instead, don't fret. You needn't bone up on this technology, because iTunes provides a way to make your videos compatible with the iPhone. Here's how: Drag an unprotected video—one that isn't a copy-protected track TV show or video you haven't purchased from the iTunes Store—onto the Library entry in iTunes' Source list.

If the video is compatible with iTunes, it will be added to the library; if not, the dragged icon will zip back to its original location. If the video isn't compatible, you can convert it with a utility such as Roxio's $40 Video Crunch for Windows (www.roxio.com) or the Mac-compatible $32 VisualHub from Techspansion (www.techspansion.com).

Some videos that play in iTunes may be encoded at resolutions or data rates too high for the iPhone to handle. Those files will not sync with your iPhone, but you can make them compatible with the iPhone. To do that, select a video (found in the Movies or TV Shows entry within iTunes' Source list) and choose Advanced > Convert Selection for iPod (**Figure 6.16**). This command creates an iPod-compatible (and, therefore, iPhone-compatible) version of the video, which you can sync to your iPhone.

Figure 6.16 Convert a video for iPod (and iPhone) compatibility.

tip **Converting a video for iPod compatibility doesn't replace the original, so it's not a bad idea to rename the converted version—*Casablanca (iPhone)*, for example—so that you can identify and sync the right one.**

Choosing and playing videos

Playing videos within the iPod area is straightforward. Tap the Videos icon at the bottom of the screen, and you'll see your videos listed by categories: Movies, TV Shows, Music Videos, and Podcasts (**Figure 6.17**). Each video has a thumbnail image of its artwork next to it. Depending on the original source of the video, you may see title, artist, season, and episode information. You'll definitely see the length of the video—1:56:26, for example.

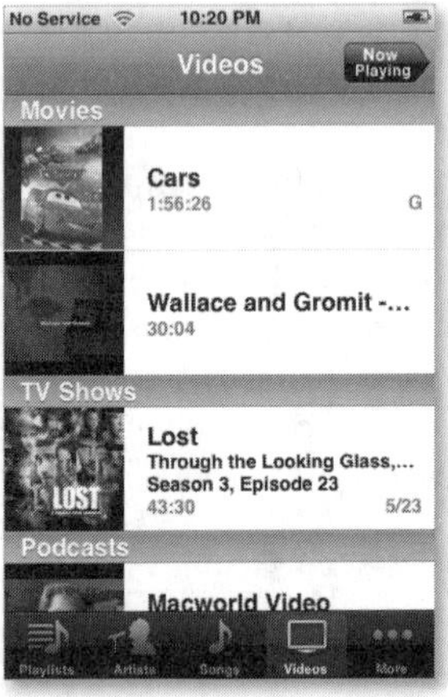

Figure 6.17 The Videos screen.

To play a video, tap its list entry. Videos play only in landscape orientation, regardless of which way you have the iPhone turned. And unlike in Cover Flow view (in which Landscape view can be obtained by orienting the Home button on the left or right), videos are always oriented so that they're right side up when your iPhone's Home button is on the right.

The video Play screen is similar to the music Play screen except that the play controls and timeline are not visible unless you make them so. To display these controls, tap the video (**Figure 6.18**). The usual play controls—Previous/Rewind, Play/Pause, and Next/Fast Forward icons, and a volume slider—appear, as does a timeline near the top of the screen. The volume slider and timeline work just like they do in the music Play screen. Drag the volume slider's volume indicator (represented by the silver dot) to increase or decrease volume and move the timeline's playhead to a new location in the video.

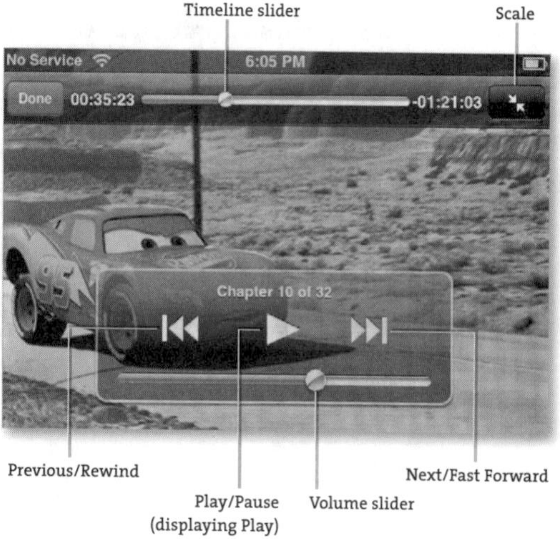

Figure 6.18 The video Play screen.

tip **Cool as these controls look, you don't need to pull them up every time you want to adjust the volume. Just use the iPhone's mechanical volume buttons.**

As I mentioned earlier in this chapter, you can advance to the next chapter in a video by tapping the Next/Fast Forward icon. (If the video has no chapters, nothing happens when you tap this icon.) If you tap and hold, the video speeds up. To retreat a chapter, tap Previous/Rewind twice (tap once to return to the beginning of the currently playing chapter). The play controls list the chapter you're currently watching—Chapter 13 of 32, for example.

The video Play screen has a control you haven't seen before: the Scale icon, in the top-right corner of the screen. Tapping this icon toggles the display between Fill Screen and Fit to Screen. (You can also toggle these views by double-tapping the display.)

Fill Screen is similar to DVDs you've seen that say the movie was altered to fit your TV. The iPhone's entire screen is taken up by video, but some of the content may be chopped off in the process.

Fit to Screen displays the entire video, similar to a letterbox movie you may have seen. In this view, you may see black bars at the top and bottom or on the sides.

When you finish watching a video, tap the screen and then tap the Done icon in the top-left corner of the screen. You'll return to the Videos screen.

tip **The iPhone remembers where you last left off. When you next play this video, it will take up from the point where you stopped playback.**

iPod settings

Like other applications and areas of the iPhone, the iPod function gets its own little entry in the iPhone's Settings screen. Tap Settings and then iPod, and **Figure 6.19** is what you see.

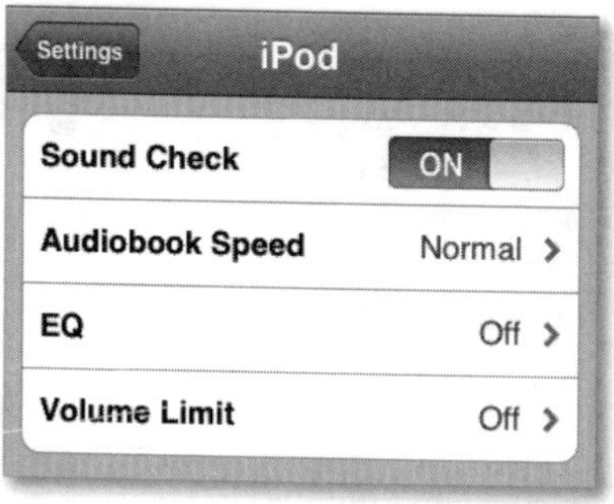

Figure 6.19 The iPod Settings screen.

Sound Check

iTunes includes a Sound Check feature that you use to make the volumes of all your tracks similar. Without Sound Check, you may be listening to a Chopin prelude at a lovely, lilting volume and be scared out of your socks when the next track, AC/DC's "Highway to Hell," blasts through your brain. With Sound Check on, each track should be closer to the same relative volume.

The iPhone includes an On/Off Sound Check option, but it works only if you've first switched Sound Check on in iTunes; iTunes must evaluate your tracks and set an instruction in each track so that it works with Sound Check. To enable Sound Check in iTunes, open iTunes' Preferences window, click the Playback icon, and check the Sound Check box. Now when you sync your tracks with the iPhone and switch Sound Check on in the iPhone's iPod Settings screen, you'll experience all that is Sound Check.

Audiobook Speed

Like a real iPod, the iPhone allows you to play tracks designated as audiobooks at Slower, Normal, or Faster speed. If you have a hard time understanding what the narrator is saying, try Slower. If you're in a hurry, give Faster a shot.

EQ

EQ (or *equalization*) is the process of boosting or cutting certain frequencies in the audio spectrum—making the low frequencies louder and the high frequencies quieter, for example. If you've ever adjusted the bass and treble controls on your home or car stereo, you get the idea.

The iPhone comes with the same EQ settings as iTunes:

- Off
- Acoustic
- Bass Booster
- Bass Reducer
- Classical
- Dance
- Deep
- Electronic
- Flat
- Hip Hop
- Jazz
- Latin
- Loudness
- Lounge
- Piano
- Pop
- R & B
- Rock
- Small Speakers
- Spoken Word
- Treble Booster
- Treble Reducer
- Vocal Booster

Although you can listen to each EQ setting to get an idea of what it does, you may find it easier to open iTunes; choose View > Show Equalizer; and then, in the resulting Equalizer window, choose the various EQ settings from the window's pop-up menu. The equalizer's ten band sliders will show you which frequencies have been boosted and which have been cut. Any slider that appears above the 0 dB line indicates a frequency that has been boosted. Conversely, sliders that appear below 0 dB indicate frequencies that have been cut.

EQ and the iPhone

Having EQ built into iTunes and the iPhone is great, but the interaction between iTunes and the iPhone in regard to EQ is a little confusing. Allow me to end that confusion.

In iTunes, you can assign an EQ setting to a song individually by clicking the song, pressing Command-I (Mac) or Ctrl-I (Windows), clicking the Options tab, and then choosing an EQ setting from the Equalizer Preset menu. When you move songs to your iPhone, these EQ settings move right along with them, but you won't be able to use them unless you configure the iPhone correctly.

If, for example, you have EQ switched off on the iPhone, songs that have assigned EQ presets won't play with those settings. Instead, your songs will play without the benefit of EQ. If you set the iPhone's EQ to Flat, the EQ setting that you preset in iTunes will play on the iPhone. If you select one of the other EQ settings on the iPhone (Latin or Electronic, for example), songs without EQ presets assigned in iTunes will use the iPhone EQ setting. Songs with EQ settings assigned in iTunes will use the iTunes setting.

If you'd like to hear how a particular song sounds on your iPhone with a different EQ setting, start playing the song on the iPhone, tap the Home button, tap Settings, tap iPod, tap EQ, and then select one of the EQ settings. The song will immediately take on the EQ setting you've chosen, but this setting won't stick on subsequent playback. If you want to change the song's EQ more permanently, you must do so in iTunes.

Volume Limit

Though Apple takes pains to warn you in the iPhone's documentation that blasting music into your ears at full volume can lead to hearing loss, some people just can't get enough volume. If your child is one of those people, consider setting a volume limit for the iPhone's headphone port. To do so, tap Volume Limit in the iPod Settings screen, and in the resulting Volume Limit screen, use the volume slider to set an acceptable volume. (Have a track playing when you do this so that you can listen to the effect.) To keep your kid from changing your settings, tap Lock Volume Limit. You'll see a Set Code screen, where you'll enter and confirm a four-digit security code (**Figure 6.20**). When this code is set, the Lock Volume Limit icon changes to Unlock Volume Limit. Tap this button, and you'll be prompted for the security code.

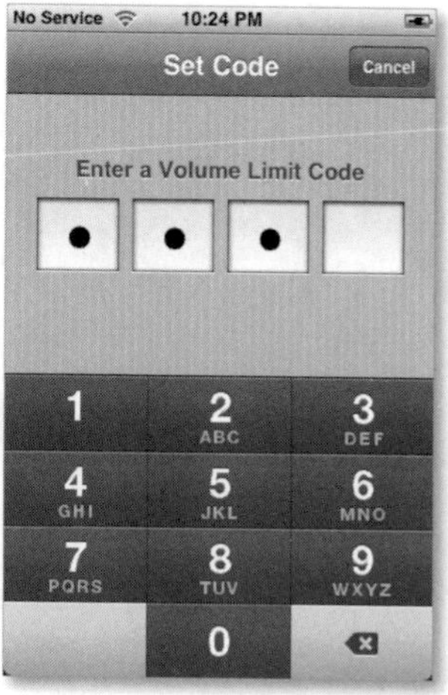

Figure 6.20 Enter and confirm a four-digit code to limit the iPhone's volume.

7

Photo, Camera, and YouTube

Your iPhone is an audio wonder, handling both calls and music, but it's also a visual delight. And no, I'm not referring to its lustrous design. I'm talking pictures—both those you take and those you view—as well as the moving pictures streamed from the free Web-based video sharing service YouTube.

The grumps among us might grouse that these features are some of the iPhone's least necessary aspects—after all, few of us absolutely *require* a phone that can display gorgeous slideshows or stream videos of piano-playing cats—but they are certainly among its most enjoyable.

In this chapter, I turn to the visual: the iPhone's photo, camera, and YouTube capabilities.

You Ought to Be in Pictures

Tapping lucky icon number 3 in the iPhone's Home screen—the one marked *Photos*—is the digital equivalent of flipping open your wallet to reveal a seemingly endless stream of pictures of the kids, the dog, and that recent trip to Pago Pago. For here, you find the pictures you've taken with the iPhone's 2-megapixel camera, plus any photos you've chosen to sync from your computer.

But the iPhone's photo feature is no mere repository for pictures. Flick a finger, and you're flying from photo to photo. If you have a more formal presentation in mind—a showing of your child's first birthday party for Grandma and Grandpa, for example—you can create something far grander in the form of a slideshow. To learn about these and other photographic wonders, just follow along.

The face of Photos

When you tap Photos, you see the Photo Albums screen, which acts as the gateway to the images stored on your camera (**Figure 7.1**). In this screen, you'll find a couple of entries (and, I hope, eventually more than just a couple).

Figure 7.1 The Photo Albums screen.

The first entry is Camera Roll. Tap it to see the images you've captured with the iPhone's camera. To the left of this entry in the Photo Albums screen, you'll see a thumbnail image of the last picture taken by the camera. To the right of the entry, in parentheses, you'll see the number of images this album contains—*Camera Roll (17)*, for example. The **>** character on the far-right edge of the screen indicates that when you tap this entry, you'll be taken to another screen. That other screen, called Camera Roll, contains thumbnail images of all the photos in this album.

The next entry, Photo Library, contains all the photos on your iPhone save for those in the Camera Roll library. It too bears a thumbnail (not one of your images, but a sunflower), and it displays the total number of images in the library—*Photo Library (584)*, for example. Tap this entry, and in the resulting Photo Library screen, you'll see thumbnail images of all the photos on your iPhone (again, excluding the Camera Roll photos).

As you learned in Chapter 2, you can sync photo albums created by such programs as iPhoto, Aperture, Photoshop Elements, and Photoshop Albums. When you do, these albums appear in the Photo Albums screen as separate entries, each featuring a thumbnail of the first image in the album as well as the number of images in the album—*Father's Day (48)* or *Family Holiday (92)*, for example. When you select your Pictures folder (Mac), My Pictures (PC), or a folder of your choosing within iTunes' Photos tab, any folders contained within those folders are presented as separate albums. So, for example, if your My Pictures folder holds three folders that contain pictures—Betty's Birthday, Dog Polisher, and Cheeses Loved and Lost—each of those items appears as a separate album in the Photo Albums screen. Again, each album lists the number of images it contains in parentheses.

Picture viewing

Think movies and TV shows look good on your iPhone? Wait until you see photos. The iPhone's bright display is the perfect portable platform for showing off your favorite photos.

As I mention earlier in this chapter, when you're in an album's screen, you see all the pictures in that album arrayed four-across as thumbnail images

(**Figure 7.2**). You can see 20 complete thumbnails on the screen. If your album contains more than 20 images, just flick your finger up across the display to scroll more images into view. To see a picture full-screen, just tap it.

Figure 7.2 A photo album's thumbnail images.

Orienteering

For those of you keeping score at home, Photos is one of those areas of the iPhone that work in both portrait and landscape orientation. If you have a picture you've taken in a camera's "normal" portrait orientation, your best bet is to flip the iPhone to landscape view (**Figure 7.3** on the next page). All the glorious wide-screen goodness of your picture will be revealed, without a large black bar slapped on either side of your photo. Fond of doing as the pros do and flipping your camera on its side for portraits? Your iPhone can accommodate these photos best when you hold the phone in its normal up-and-down orientation.

Figure 7.3 Widescreen picture view.

The iPhone will automatically rotate and resize your images to accommodate the phone's orientation. And unlike videos, which display on the horizontal only when the Home icon is on the right side of the screen, photos will display in their correct orientation regardless of which way you've turned the iPhone—up, down, left, or right.

tip **If you always want to view your photos at their best advantage—horizontal shots viewed in landscape orientation and vertical pictures in portrait orientation—yet you tire of constantly flipping the phone, try this: Create photo albums based on picture orientation. Put all your portrait-oriented photos in one album and your favorite landscape shots in another. When it's time to view them, they'll always look their best without your having to reorient the phone.**

The picture screen

In addition to letting you rotate your pictures by flipping your phone around, the screen in which you view individual images offers some other cool features. When viewing a picture in an album or the Photo Albums screen, you'll briefly see a transparent gray control bar at the bottom of the screen that displays four symbols: Options, Previous, Play, and Next (**Figure 7.4**). When viewing a picture in the Camera Roll album, you'll see one additional icon: Trash. If you've taken a picture that you now regret, just tap the Trash icon, and in the sheet that appears, tap the red Delete Photo icon.

Figure 7.4
The picture screen.

This control bar conveniently disappears after a couple of seconds so you can see the complete picture without obstruction. To bring it back, just tap the display.

The left and right arrow icons that represent the Previous and Next commands do just what they suggest. Tap the left-pointing arrow, and you move to the previous image in the album. Tap the right arrow, and you're on to the next image. If you tap and hold these icons, you'll zip through your pictures at increasing speed. When you press the Play icon, a slideshow begins, starting with the current image and playing to the end.

The timing and transitions of your slideshow are determined by options for the Photos entry in the iPhone's Settings screen. You have the option to play each slide for 2, 3, 5, 10, or 20 seconds. And you can choose among Cube, Dissolve, Ripple, Wipe Across, and Wipe Down transitions.

When you tap the screen during a slideshow, another transparent gray bar appears briefly at the top of the screen. This one displays a left-pointing arrow bearing the name of the currently selected photo album. As with most iPhone screens, you tap this arrow to move up one screen in the iPhone's hierarchy. You'll also see an entry such as *8 of 48*, which tells you which one of the total number of pictures you're looking at.

Moving pictures

Tapping those Previous and Next icons is the less impressive way to move from picture to picture. For a far more stirring demonstration of the iPhone's slickness, swipe your finger to the left to advance to the next picture or to the right to retreat back one picture. You're guaranteed to get an "Oooh!" from the audience on this one.

While you've got your audience in the "Oooh"ing mood, try this: Double-tap a picture. Like magic, the screen zooms in on the center of the picture. Drag your finger on the picture to reposition it. If you'd like greater control over how large the image is, use the spread gesture to make it grow incrementally. Regrettably, the iPhone won't remember how you've repositioned and resized the picture. You also can't swipe to the next picture until you've restored the picture to its original size. You can do this by double-tapping the display again or by pinching the image down to its native size.

Swiping is good at any time, even during a slide-show. If, while viewing a slideshow, you'd like to take control, just tap the display to stop the slideshow, or swipe your finger to the left to advance or right to go back. When you manually navigate to the photo that precedes or follows the one on view, the slideshow is canceled. To restart it, you must tap the display to produce the Play controls and then tap the Play icon.

These settings are the defaults. If you've configured Photo Settings so that the Repeat and Shuffle options are on, the slideshow will behave a bit differently. To begin with, the show will reach the end and then start over, continuing to play until you tell it to stop by tapping the display. And if Shuffle is on, the photos in the selected album will play in random order.

The Options icon

No, I haven't forgotten about this icon, which resembles an arrow emerging up out of a tiny photo. The features it provides are enough to warrant their own little section.

When you tap Options, a pane scrolls up from the bottom of the iPhone screen, displaying the entries Use As Wallpaper, Email Photo, Assign To Contact, and Cancel (**Figure 7.5**). Mac users who subscribe to Apple's .Mac Web service may also see a fifth entry, labeled Send to Web Gallery.

Figure 7.5 Options available in the Picture screen.

These commands work this way:

Use As Wallpaper. Tap this command, and the iPhone will offer to use the picture on display as your iPhone's wallpaper—the image that appears when your phone is locked or when you're speaking with a contact who doesn't have an associated picture. You're welcome to move the picture around by dragging your finger on the screen. You can also scale it larger by using the spread gesture. When you're happy with the picture's position, tap Set Wallpaper.

Email Photo. Tap this command, and the picture on view is placed in an empty email message. You're welcome to tap just above the picture in the message body and add text to the message. Fill in the To and Subject fields, and press Send. Your message-with-picture-attachment is whisked away.

note **If you have multiple email accounts set up in iPhone Mail, you can't choose which address to send the message from; it'll be sent via the default account you've entered in Mail Settings. Also, after you've sent the message (or tapped Cancel to forget the whole thing), you'll return to the picture screen rather than to the Mail application.**

Assign To Contact. As the name suggests, this command allows you to associate the picture on view with a contact of your choosing. Tap the command, and up scrolls your contact list. Use the usual finger-drag or tap-a-letter method to navigate to the contact you want; then tap the contact, and a screen similar to the Use As Wallpaper screen appears. Move and scale the image to your liking, and tap Set Photo to assign the picture to the contact. When you speak to that contact on your iPhone, the picture you've selected will appear.

tip

You can use pictures to express your true feelings about a contact. If your boss earned his position by being first cousin to the company president, for example, sync a picture of Bozo to your iPhone, and assign it to the clown who controls you.

Send to Web Gallery. If you're a Mac user with a .Mac account (Apple's $100-per-year subscription package of online services) and a copy of iLife '08, you can use this icon to upload pictures directly from your iPhone to a .Mac Web Gallery—a special .Mac Web site that hosts pictures as well as movies. Note that this icon appears only if you've added a .Mac account to Mail. If you don't have a .Mac account, or if you haven't added your .Mac account to Mail, you won't see this button.

Snapping Pictures

The iPhone contains a 2-megapixel camera capable of capturing JPEG images at a resolution of 1600 x 1200 pixels. Its lens is in the top-left corner of the back of the iPhone (and yes, this position does make it easy to plant your pinkie in front of it accidentally). It takes perfectly decent pictures in well-lit environments. In low-light settings—an indoor concert, outside at night, or in a poorly lit room—the results are less than stellar. The iPhone has no flash, and you can't zoom the camera, so it's the ultimate point-and-shoot.

Taking a photo

To use the camera, simply tap the Camera icon on the iPhone's Home screen. In a short time, you'll see the image of a closed shutter and then a view of whatever is in front of the camera lens. To take a picture, tap the camera button at the bottom of the screen (**Figure 7.6**). Its little icon will show you the camera's orientation: portrait or landscape.

Figure 7.6 Tap the Camera button to take a picture.

After you snap a picture, its image collapses into a Pages icon in the bottom-left corner of the iPhone screen, indicating that the image has been saved. Tap this icon, and you're taken to the Camera Roll screen, which displays the camera's captured images as thumbnails. As in the Photos application, you can tap a thumbnail for a full-screen view of the picture. To return to the camera, tap the Pages icon again.

In Camera Roll full-screen view, you have the same controls that you have in the Photos Play screen, plus the Trash icon, which you tap to discard the picture currently on screen. And yes, the Options icon appears here, too. Tap it, and you'll find the expected Use As Wallpaper, Email Photo, Assign To Contact, and, possibly, Send to Web Gallery buttons.

In the top gray bar of this screen, you'll see a Camera icon. To return to the camera, just tap this icon.

Syncing camera photos to your computer

Wonderful as the iPhone may be for showing your photos, suppose that you take a really great picture with it that you'd like to print. Somehow, you've got to get it out of your iPhone and onto a computer.

"Ooh, ooh, ooh!" you're calling anxiously from the back of the room. "I know! I know!! *I KNOW!!!* Just tap Options and email the picture to yourself!"

Great idea!

But ... no. Although you get big points for remembering what I told you in the past few pages, the correct answer is: Plug your iPhone into your computer. When you do, in all likelihood, your computer will automatically offer to copy the pictures from the iPhone's Camera Roll album to your computer (**Figure 7.7**).

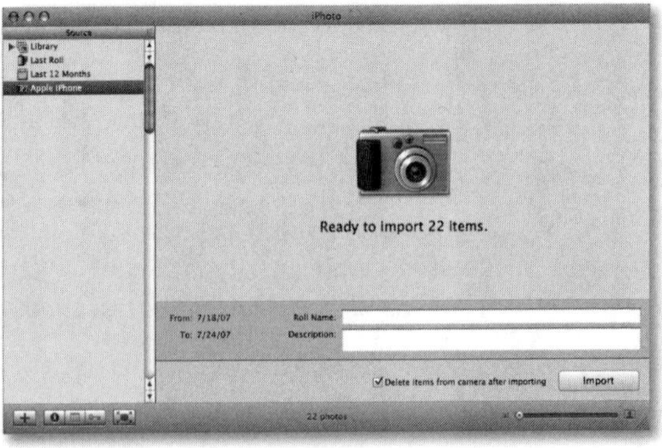

Figure 7.7 By default on a Mac, iPhoto launches when you plug in a picture-laden iPhone and offers to copy its camera photos.

Although you could email the pictures to yourself, when you do, you send a picture that is of lower resolution than the original. Specifically, photos emailed from the iPhone are sent at a resolution of 640 x 480 pixels. When you download the pictures directly to your computer, you get full 1600 x 1200 images.

When you sync a Mac with an iPhone containing photos taken with its camera , by default, iPhoto launches and asks whether you'd like to add the pictures from your iPhone to its library. On a Windows PC, the AutoPlay option will launch, listing options including importing pictures from your "camera." Do this, and the pictures are copied from the iPhone to your computer.

tip

Some Mac users are driven to distraction by this behavior. They'd like to import pictures when *they* want to, not every time they sync their phones. To stop this from happening, go to your Applications folder, launch Image Capture, select Image Capture > Preferences, click the General tab, and choose No Application from the When a Camera Is Connected Open pop-up menu. From now on, when you want to pull pictures off the camera, you must launch iPhoto. When you do, iPhoto will see your iPhone and offer to copy its pictures.

When you copy those photos, they're not removed from your iPhone automatically. Both iPhoto and Aperture provide an option to copy the pictures and then erase the card or camera where the original images are stored. This option applies to your iPhone. Enable it, and after your pictures are copied to your computer, the photos are deleted from the iPhone.

YouTube

The iPhone's ability to stream content from YouTube, the Internet video sharing service, came as a surprise in the few weeks just before the iPhone's release. Up until that point, we knew that the iPhone could play videos synced to it through iTunes. But streaming video? Who knew? Well, Apple and YouTube, apparently, but they weren't about to tell us.

Tap the YouTube icon on the iPhone's Home screen, and you'll see a screen that resembles the one you view when you enter the iPhone's iPod area. Like the iPod screen, this one has five icons along the bottom. By default, these icons are Featured, Most Viewed, Bookmarks, Search, and the ever-popular More (**Figure 7.8**).

Figure 7.8 Icons in the YouTube screen.

Here's what you'll find when you tap each icon.

Featured

Tap Featured, and you get a list of the 25 YouTube videos that the service believes most worthy of your attention (**Figure 7.9**). To play one, just tap it. The video will stream to your iPhone either via a Wi-Fi connection (if one's available) or over EDGE. Naturally, Wi-Fi will bring the video to you faster. When you scroll to the bottom of the list, you'll see a Load 25 More entry. Tap it, and another 25 videos are added to the list.

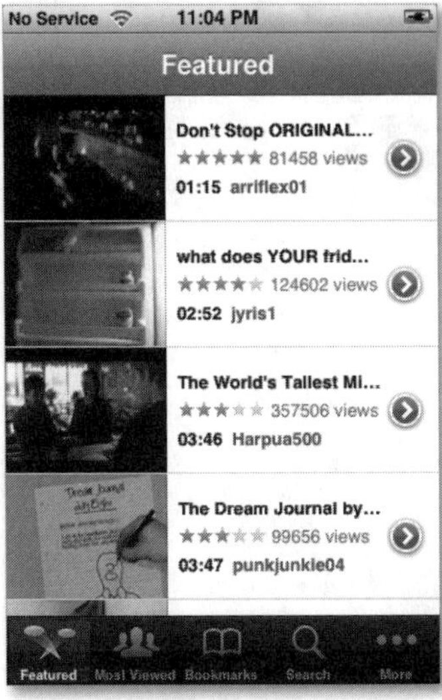

Figure 7.9 Featured YouTube videos.

If a video's title, such as *Nora The Piano Cat,* doesn't provide you enough information, feel free to tap the blue icon to the right of the video's title. When you do, you'll see an information screen that tells you the date when the video was added; its category (Drama or Documentary, for example); and its tags, which include anything the poster thought appropriate, such as *poodle*, *waterslide*, and *ointment* (**Figure 7.10** on the next page).

If you'd like to bookmark or share the video, tap the appropriate icon in this screen. Tap Bookmarks, and the video is added to your list of YouTube bookmarks, allowing you to retrieve it easily another time. When you tap Share, a new email message opens. The Subject line includes the title of the video, and

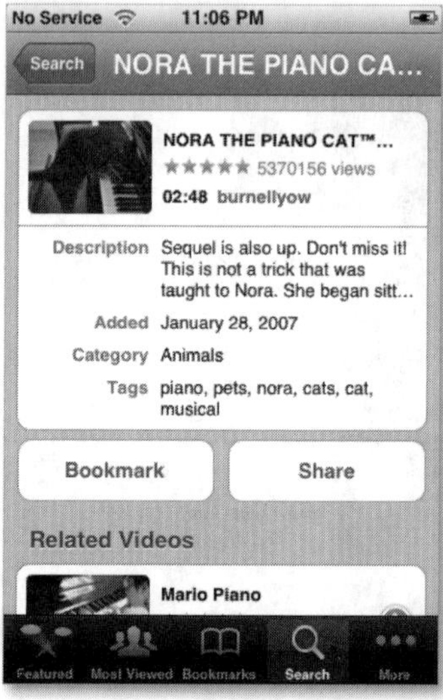

Figure 7.10 A YouTube video information screen.

the message body contains *Check out this video on YouTube:*, followed by a link to the video. (You can edit *Check out this video on YouTube:* to anything you like.) When you complete the To field and tap Send, the email message is sent via your default email account (as configured in Mail Settings).

The description screen also includes a Related Videos area. If YouTube has videos that it believes are similar in theme to the one you've chosen, it will list them here.

Most Viewed

The Most Viewed icon provides you the opportunity to view YouTube's most popular videos—all videos, or the most viewed today or this week. Like the Featured screen, this one carries a Load 25 More entry at the bottom of the list. To determine whether you watch all, today, or this week's most viewed videos, tap the appropriate icon at the top of the screen.

Bookmarks

As the name hints, here is where your YouTube bookmarks are stored. (The videos themselves aren't stored on the iPhone—just the links to them.) To begin streaming one of these videos, just tap its name. To remove a bookmark, tap the Edit icon at the top of the screen, tap the red minus sign (–) that appears next to the entry, and then tap Delete (**Figure 7.11**).

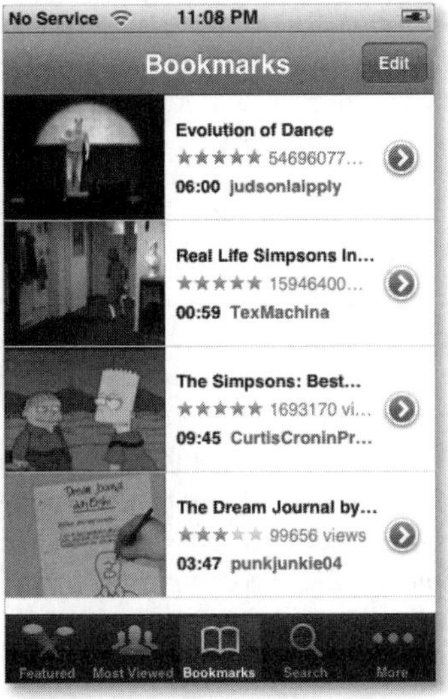

Figure 7.11 YouTube bookmarks.

When you're finished removing bookmarks, tap Done, and you'll return to the Bookmarks screen.

note **These bookmarks apply only to those videos you've bookmarked on your iPhone. Yes, you may have bookmarked YouTube videos on your Mac or PC, but no, you can't transfer these bookmarks to your iPhone.**

Search

You can search YouTube's catalog of videos, and of course, this is the way to go about it. Tap Search, and you get a Search field in return. Tap this field, and up pops the iPhone's keyboard. Type a search term—*skateboard* or *Mentos,* for example—and YouTube searches for videos that match your query. Then the service presents a list of 25 videos that it feels match what you're after. If more than 25 videos are available that match your query, your friend the Load 25 More entry appears at the bottom of the list.

note **If you purchased this book shortly after its release, you may be disappointed that your iPhone is unable to find as many matches as the "real" YouTube on your computer. The reason is that the "real" YouTube encoded content in Adobe Flash format—a format that the iPhone can't play. As I write this chapter, YouTube is in the process of converting its content to the iPhone-friendly H.264 format. This change is a boon not only for iPhone users, but also for all YouTube fans, because video looks better encoded this way. YouTube expects to complete the conversion process by autumn 2007.**

More

You've read Chapter 6, right? Then this More icon should be no mystery to you. Tap it, and you're presented three additional choices: Most Recent, Top Rated, and History. Most Recent offers a glimpse of the 25 videos most recently added by YouTube. Top Rated displays YouTube's 25 highest-rated videos. And History details all the videos you've chosen. Yes, *chosen*. You don't have to watch these videos in order for them to appear in your History list. Just choose them, and even if you cancel them before they appear, they'll be part of your iPhone's YouTube History. If this list is getting too long, or if you're simply embarrassed by some of the things you've chosen, tap the red Clear icon at the top of the screen. All History entries will disappear.

note **The Clear icon is an all-or-nothing affair. Currently, the iPhone doesn't provide an option to delete individual videos from the History screen.**

Playing YouTube videos

To play a YouTube video, tap it, and the video will begin loading in landscape orientation. You'll see the now-familiar video play controls—Back, Play, and Forward—along with a volume slider, timeline, and fill-screen icon. Like the play controls in the iPhone's iPod area, these controls fade a few seconds after they first appear. To force them to reappear, just tap the iPhone's display.

In addition to the play controls, you'll see a Bookmarks icon to the left of the play controls and a Share icon to the right (**Figure 7.12**). Tap Bookmarks, and the currently playing video is added to your YouTube bookmarks. Tap Share, and you create another one of those special YouTube recommendation emails that I describe in the "Featured" section earlier in this chapter.

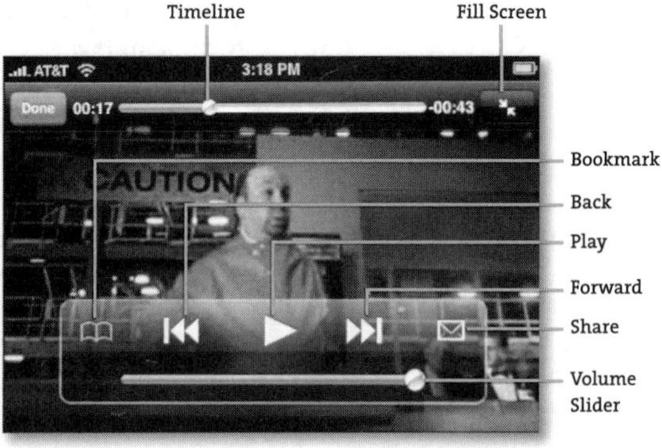

Figure 7.12 The YouTube play screen.

The video will begin playing when the iPhone determines that it has downloaded enough data so that the video will play from beginning to end without pausing to download more. When the video concludes, you'll see a description screen for it.

8

Stocks, Map, Weather, Clock, Calculator, & Notes

I've covered the iPhone's major areas—Phone, Mail, Safari, and iPod—as well as its most significant applications, including SMS, Calendar, Photos, Camera, and YouTube. It's time to turn to the smaller applications, which occupy the 6th through 11th positions on the iPhone's Home screen: Stocks, Maps, Weather, Clock, Calculator, and Notes.

If you've used Mac OS X, many of these applications will be familiar to you, as most of them are offered in that operating system as *widgets*—small applications that perform limited tasks. On the iPhone, they're considered to be full-blown applications, even though they're largely single-purpose programs. They work this way.

Stocks

The Stocks application is nearly a perfect clone of the Mac OS X Stocks widget. Like that widget, the application displays the stocks and market indexes (Dow Jones Industrial Average and NASDAQ, for example) you choose in the top part of the screen and performance statistics below. Next to each index or stock ticker symbol, you'll see the almost-current share price (results are delayed by 15 minutes)—*AAPL, 143.85*, for example—followed by the day's gain or loss, as represented by a green (gain) or red (loss) icon.

By default, the application represents gains and losses in points—*+3.89*, for example. You can toggle to a percentage view and change back to points by tapping one of these red or green icons.

note **You must be connected to the Internet in some way—either via ATT's EDGE network or a Wi-Fi connection—for the results to appear.**

To view statistics for a particular index or stock, just tap its name. A graph at the bottom of the screen charts that index's or stock's performance over 1 day, 1 week, 1 month, 3 months, 6 months, 1 year, or 2 years (**Figure 8.1**). To choose a time period, just tap the appropriate icon (1d for 1 day or 6m for 6 months, for example).

Figure 8.1 Stocks application.

For more detailed information on an index or stock, tap its name to highlight it and then tap the tiny Y! (for Yahoo) icon in the bottom-left corner of the screen. Doing so launches Safari and whisks you to a Yahoo oneSearch page with links related to that item. There, you'll find links to a Yahoo Finance page devoted to the index or stock, with related news, products, full and mobile Web pages, and Web images.

If you tap the *i* (Information) icon in the bottom-right corner of the display, the screen flips to reveal the indexes and stocks that appear on the application's front page. Click the plus (+) icon in the top-left corner and use the iPhone's keyboard to add a ticker symbol or company name. (In the case of a company name, the iPhone will search for matches. If you type **Apple** and tap Search, for example, you'll get a list that includes not only Apple, Inc., but also MS Apple Sparqs and Golden Apple Oil and Gas.) Tap the search result you want, and it will be added to the bottom of the list. Regrettably, there's no way to reorder this list. To remove items, just tap the red minus (–) symbol next to the item's name and then tap the resulting Delete icon.

The Information screen also includes two icons: % and Numbers. Tap one or the other to determine how gains and losses are denoted on the main Stocks screen.

This screen also offers a less-obvious icon. To have Safari take you to the Yahoo Finance page, simply tap Yahoo! Finance at the bottom of the screen. When you're ready to flip back to the main Stocks screen, tap Done in the top-right corner of the Information screen.

Maps

This feature is a version of Google Maps made for the iPhone, and I've found it to be one of the phone's most useful applications. You can use it to search for interesting locations (including businesses, residences, parks, and landmarks) and give those locales a call with a couple of taps. You can also get driving directions between here and there—and in some cases, check traffic conditions along your route.

The Maps application has two major components: Search and Driving Directions. Each is available from the main Maps screen.

Search and explore

At the top of the Maps screen, you see a Search field (**Figure 8.2**). Tap it, and up pops the iPhone's keyboard. With that keyboard, you can enter any of a variety of search queries, including contacts in your iPhone's address book (*Joe Blow*), a business name (*Apple, Inc.*), a town name (*Springfield*), a more-specific town name (*Springfield, MO*), a street or highway name (*Route 66*), a specific street name in a particular town (*Main Street, Springfield, MO*), or a thing (*Beer*).

Figure 8.2
Maps' search feature.

If Maps is confused by your search, it will offer a list of possibilities in a Did You Mean dialog box. When typing **Main Springfield MO**, for example, you'll be asked to choose between *N Main Ave, Springfield, Greene, Missouri* and *S Main Ave, Springfield, Greene, Missouri*. You can help end some of that confusion by entering a more specific search, such as **Main, Springfield, MO 65802** or **Beer 95521**. In short, the more specific you are in your query, the more accurate Maps will be.

Maps' search results can be displayed in three views:

- Maps, which is a graphical illustration of the area
- Satellite, which is a photo captured by an orbiting satellite
- List, which is a ... well, *list* of all the locations pinpointed on the current map

These options are available as labeled buttons at the bottom of the screen.

In Maps and Satellite views, search results are denoted by red push pins that drop onto the map. Tap one of these pins, and the name of the item appears. Again, this name can be the name of a contact, business, town, or highway. List items are accompanied by blue › icons that you tap to take you to that location's Info screen.

Info screens present any useful information Maps can obtain about an item, including phone number, email, address, and home-page URL (**Figure 8.3**). The phone number, email, and URL links are *live*, meaning that when you tap a phone number, your iPhone will place a call to the number; when you tap an email address, Mail opens and addresses a message to that contact; and when you tap a URL, Safari opens and displays that Web site.

Figure 8.3 Maps' Info screen.

At the bottom of each Info screen, you'll see five labeled buttons: Directions To Here, Directions From Here, Add to Bookmarks, Create New Contact, and Add to Existing Contact. You may have to scroll the screen to see all these buttons. They work this way.

Directions To Here

Tap this button to display Maps' Driving Directions interface (which I'll explain shortly), with the item's address in the End field.

Directions From Here

This feature works similarly. The difference is that the item's address appears in the Driving Directions Start field.

Add to Bookmarks

You can bookmark locations in Maps. Tapping this icon brings up the Add Bookmark screen, where you can rename the bookmark, if you like. When you're done, tap Save, and that location is available from Maps' Bookmarks screen (which, again, I'll get to shortly).

Create New Contact

Having searched for, say, *Apple, Inc.*, you may want to create a contact for this place of business so you can easily call the main switchboard to tell the operator how satisfied you are with your iPhone purchase. As its name hints, tapping this icon causes a New Contact screen to appear, with the information from the Info screen filled in. You're welcome to add any other information you like, using the standard contact-field tools.

Add to Existing Contact

You say your buddy Brabanzio has just started putting in his 8 hours at the local pickle works? Use Maps to locate said works, tap this icon, and add its information to his contact information.

Bookmarks screen

The Search field includes a very helpful Bookmarks icon. Tap this icon to bring up a list of all the locations you've bookmarked, as well as recent search terms and your list of contacts (**Figure 8.4**).

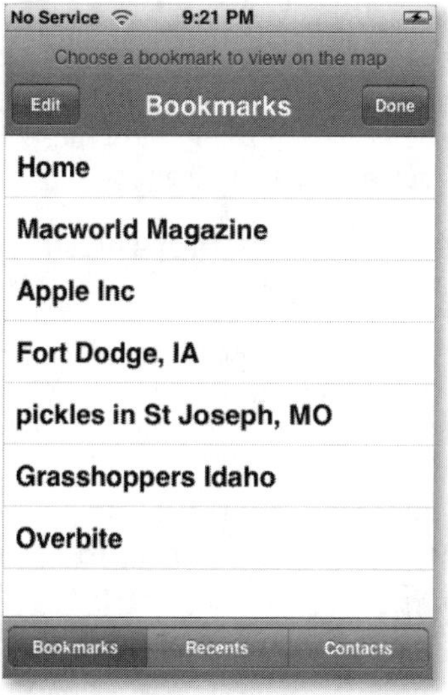

Figure 8.4 Maps' Bookmarks screen.

Bookmarks

To remove, rename, or reorder select bookmarks, tap the Edit icon. In the resulting screen, you can tap the now-expected red minus (–) icon to produce the Delete icon, which you tap to remove the bookmark. You can also tap the bookmark to show the Edit Bookmark screen, where you can edit the bookmark's name. Finally, you can reorder bookmarks by dragging the right side of the bookmark up or down the list.

Recents

Tap the Recents icon, and you'll see a list of the previous 20 searches done on your iPhone. As you conduct a new search, the last search in this list is deleted. These queries are categorized by Search (*pizza*), Location (*Grand Rapids*), and Contact (*Ebenezer Scrooge*). Tap one of these entries, and its location appears on the map.

How Does Maps Know Where I Am?

The iPhone does not include any global positioning system (GPS) hardware. Unlike the GPS unit in your brother-in-law's car, the iPhone doesn't know where it is until you tell it. So how does it know where to look for the nearest juke joint or pinball parlor? By the location of your last query.

If you last asked Maps to locate something in San Francisco, California, for example, and then entered the term **juke joint** in the search field, you'd learn about Oakland's Dotha's Juke Joint on Broadway. Enter that same term after searching in Chicago, and you'll find links to the Jazz Record Mart as well as to Chicago's House of Blues.

Because Maps bases its suggestions on the last location searched, it's not a bad idea to create a bookmark for places you often go—one for work and another for home, for example. Choose a particular bookmarked location to get you close to what you seek (a local library or pool hall, for example) and then enter a search term that zeroes in on the location you desire.

Contacts

It's swell that your Aunt Vilma has sent you a change-of-address card, but where the heck is Fort Dodge, Iowa? Tap the Contacts icon, find Aunt Vilma's name in the long list of contacts, tap her name, and then tap the street address of her new cabin down by the river. Maps will pin her palace in next to no time.

Which way do I go?

Maps' Driving Directions component is the next-best thing to having a GPS strapped to your wrist. Feed it the locations of where you'd like to start and where you'd like to end up, and it provides a reasonable route for getting there. Like so:

1. Tap the double-arrow icon in the bottom-left corner of the Maps screen.

   Empty Start and End fields appear at the top of the screen.

2. Tap the Start field to bring up the iPhone's keyboard, and type a location from which you'd like your journey to begin.

   This location can be something as generic as a zip code or as specific as your home address. Alternatively, you can tap the Bookmarks icon and then tap a bookmark in the resulting screen, and its location will appear in the Start field.

3. Tap the End field.

Same idea—type a location or choose a bookmark (**Figure 8.5**).

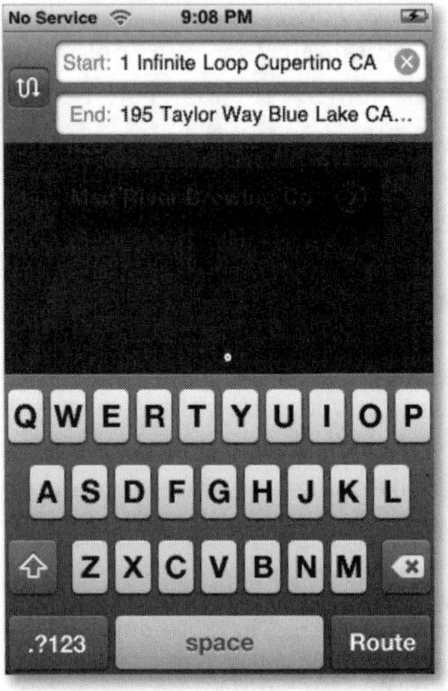

Figure 8.5 Driving directions.

4. Tap the blue Route button in the bottom-right corner of the Maps screen.

   Maps will present an overview map of your route, displayed in Map view (versus Satellite or List view). At the top of the screen, you'll see the route's distance and the estimated time of your journey (**Figure 8.6**).

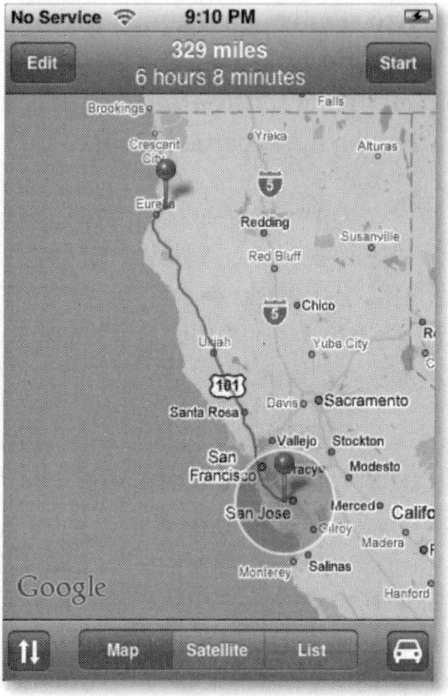

Figure 8.6
Trip overview.

5. Tap the Automobile icon in the bottom-right corner of the Maps screen.

   If the traffic area you're in supports the feature (and not all areas do), you'll see colored lines that indicate how congested the roads are. Green denotes good traffic flow, yellow is somewhat congested, and red is stop-and-go traffic (sometimes just stop). Yellow and red areas on the map throb so that they're more noticeable. If the service isn't supported in the area you're looking at, you won't see any changes in the map when you tap this icon.

tip **Be sure to zoom in on the map when you see yellow and red traffic warnings. The warning may apply only to one direction of traffic—with luck, the direction you're *not* traveling in. A zoomed-in view will tell you.**

At the bottom of the screen, you'll see the usual Map, Satellite, and List buttons. The Map and Satellite buttons do exactly what you'd expect, but List's functionality changes when you're using Driving Directions.

Tap List, and the twists and turns of your route are laid out in numbered steps—for example, ***1*** *Turn right at Old Codger Road – go 0.4 mi.* **2** *Merge onto CA-94 W via ramp – go 22.6 mi.* Tap a step, and Maps displays that portion of your trip on a map, circling the important twist or turn outlined in the step as well as displaying the written driving directions for that step at the top of the screen.

To view the next turn in your trip, just tap the right-arrow icon at the top of the screen. To return to the map overview of your trip, tap List; tap the first step in your trip; and then tap the left-arrow icon (or continue tapping the left-arrow icon until you see the overview).

If you like this turn-by-turn graphic overview of your route, you can skip the List icon altogether. From the route overview screen, just tap the Start icon in the top-right corner of the screen. The first step of your journey will be shown in all its graphic glory, along with the accompanying text placed at the top of the screen (**Figure 8.7**). Tap the right-arrow icon to proceed to the next step. Should you want to edit your route—change the Start or End point—just tap the description at the top of the display. The Start and End fields will appear, along with the iPhone's keyboard.

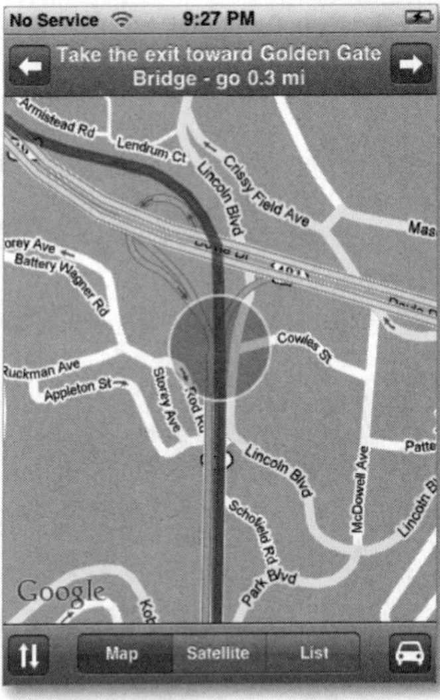

Figure 8.7
Taking a trip.

tip

If you don't want to become another gruesome statistic, please don't try to follow one of these routes while driving. Multitasking is tempting, of course, but the iPhone's print is a little small, and it's not safe to take your eyes off the road to read the next direction or tap a couple of icons. Give your iPhone to a passenger, pull over, or glance at the directions when you're at a long stoplight.

Weather

Weather is another iPhone application that owes more than a tip of the hat to a Mac OS X widget. Though the layout of the iPhone's Weather application is vertical rather than horizontal, it contains the same information as its namesake widget: a 6-day forecast (including the current day); current temperature in Fahrenheit or Celsius (selectable from the application's Information screen); each day's projected highs and lows; and icons that represent the current or projected weather conditions, such as sun, clouds, snow, or rain (**Figure 8.8**).

Figure 8.8 Weather application.

To move from one location screen to the next, simply swipe your finger horizontally across the screen. Alternatively, just tap to the right or left of the small white dots that appear at the bottom of the screen. (These dots indicate how many locations you have saved.)

Like the Stocks application, this one bears a small Y! icon in the bottom-left corner of the screen. Tap it, and sure enough, Safari displays Yahoo's oneSearch page. There, you'll find links to city guides, expanded weather information (if available), events in that location, Flickr photos and Web images associated with the location, and full- and mobile Web pages that have something to do with the location.

Tap the *i* icon in the bottom-right corner of the Weather screen, and the screen flips around to display all the locations you've saved. To add a new one, tap the plus (+) icon; use the iPhone's keyboard to enter a location (again, a zip code is a handy shortcut); and tap Search. To remove a location, tap it; tap the red minus (–) icon; and then tap Delete. To switch from Fahrenheit to centigrade, tap the appropriate icon at the bottom of the screen. Regrettably, you can't reorder locations in the Weather application.

If you'd like to take a quick trip to the Weather Channel page (powered by Yahoo), just tap the Weather Channel icon at the bottom of the Information page. When you're finished, tap Done to return to the forecast view.

Clock

More than just a simple timepiece, the iPhone's Clock application includes four components—World Clock, Alarm, Stopwatch, and Timer—that are available as icons arrayed across the bottom of the application's screen. Here's what they do.

World Clock

Just as its name implies, World Clock allows you to track time in multiple locations. Clocks are presented in both analog and digital form (**Figure 8.9**). On analog clocks, day is indicated by a white clock and night by a black one.

Figure 8.9 World Clock.

To add a new clock to the list, just tap the plus (+) icon in the top-right corner of the screen. In the

Search field of the resulting keyboard screen, enter the name of a reasonably significant city or a country. The iPhone includes a database of such cities and offers suggestions as you type.

You can remove or reorder these clocks. Tap Edit and use the red minus (–) icon to delete a clock. To reposition a clock, tap its right side and drag it up or down in the list.

Alarm

Your iPhone can get you out of bed in the morning or remind you of important events. Just tap Alarm at the bottom of the screen and then tap the plus (+) icon to add an alarm (**Figure 8.10**).

Figure 8.10 Get up with the alarm clock.

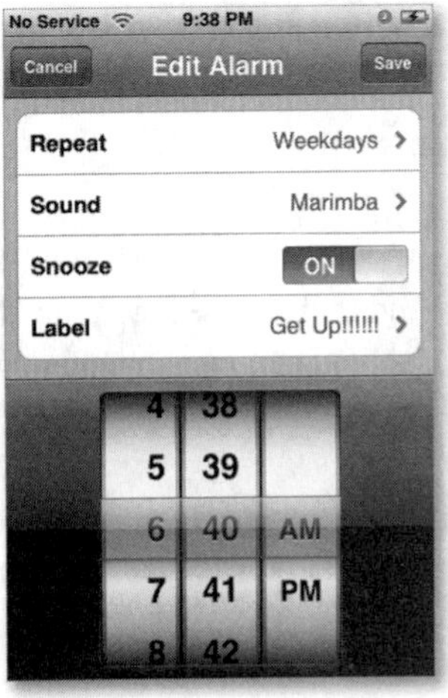

In the Add Alarm screen, you'll find a Repeat entry, which allows you to order an alarm to repeat each week on a particular day; a Sound entry, where you assign one of the iPhone's 25 ringtones to your alarm; an On/Off Snooze entry, which, when on, tells the iPhone to give you 10 more minutes of shuteye when you press the Home button; and a Label entry that allows you to assign a message to an alarm (*Get Up, Meeting This Morning* or *Drink Your Milk,* for example).

To create a new alarm, just flick the hour, minute, and AM/PM wheels to set a time for the alarm. Tap Save to save the alarm. When you save at least one alarm, a small clock icon appears in the iPhone's status bar.

tip

You can create an alarm only for the current 24-hour period. If you'd like an alarm to go off at a time later than that, use the Calendar application to create a new event, and attach an alert to that event.

Stopwatch

Similar to the iPod's Stopwatch feature, the iPhone's Stopwatch includes a timer that displays hours, seconds, and tenths of seconds. Tap Start, and the timer begins to run. Tap Stop, and the timer pauses. Tap Start again, and the timer takes up where it left off. Tap Reset, and the timer resets to 00:00.0.

While the timer runs, you can tap Lap, and a lap time will be recorded in the list below. Subsequent taps of Lap add more lap times to the list. When you tap Lap, the counter resets to 0.

Timer

The iPhone's Clock application includes a timer that will tick down from as little as 1 minute to as much as 23 hours and 59 minutes. To work the timer, just use the hour and minute wheels to select the amount of time you'd like the timer to run; then tap Start (**Figure 8.11**). (Alternatively, you can tap a number on the wheel, and it will advance to the "go" position.) The timer will display a countdown in hours, minutes, and seconds and the label on the Start button will change to Cancel. Tap Cancel to stop the countdown.

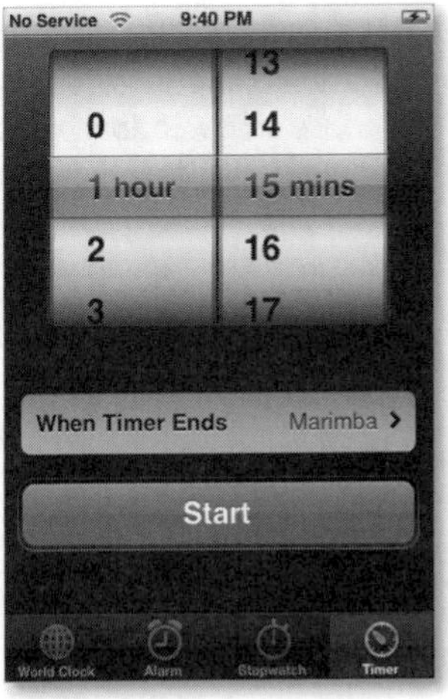

Figure 8.11 Time keeps on tickin', tickin', tickin' into the future....

The iPhone offers two actions when the timer ends: Either it will play one of its ringtones (plus vibrate and display a Timer Done dialog box), or it will sleep the iPhone's iPod feature. The latter option isn't as odd as it first sounds. Many people like to listen to soothing music or ambient sounds as they drift off to sleep. The Sleep iPod option allows them to do just that without playing the iPhone all night (and needlessly running down the battery).

Calculator

Unless you've stubbornly clung to your grandfather's abacus, you've used an electronic calculator like this before. Similar to the dime-a-dozen calculators you can find on your computer or at the local Bean Counters "R" Us, the iPhone's Calculator application (**Figure 8.12**) performs addition, subtraction, division, and multiplication operations up to nine places. When you choose an operation (addition or subtraction, for example), Calculator highlights that symbol by circling it.

In addition to the 0–9 digits and the divide, multiply, add, subtract, and equal keys, you find these additional keys:

- **C** Tap C to clear the total.
- **m+** Tap m+ to add the displayed number to the number in memory. If no number is in memory, tapping m+ stores the displayed number in memory.
- **m–** Tap m– to subtract the displayed number from the memorized number.
- **mr/mc** Tap mr/mc once, and the displayed number replaces the currently memorized number.

Tap mr/mc twice, and memory is cleared. A white ring will appear around this key if a number is in memory.

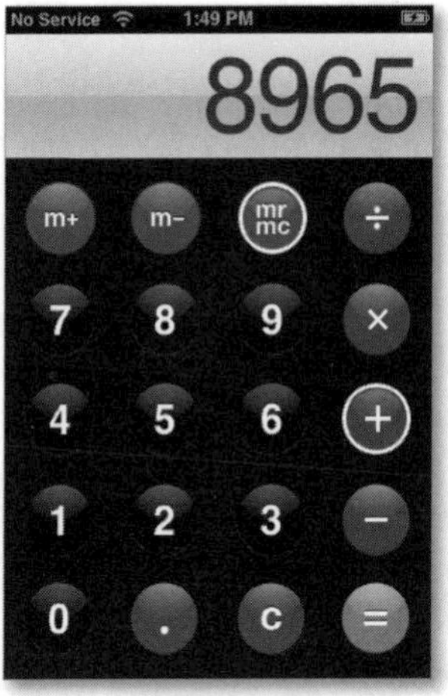

Figure 8.12 Calculator application.

Notes

Notes is the iPhone's simple text editor—and by *simple*, I mean downright rudimentary. Tap Notes on the iPhone's Home screen and then tap the Plus icon in the top-right corner of the resulting Notes screen to create a new note. When you do, the iPhone's familiar keyboard appears. Start typing your new novel (OK, novelette). If you make a mistake, use the usual text-editing tricks to repair your work.

Each individual Notes screen has four icons at the bottom. The left-arrow and right-arrow icons do exactly what you'd expect: move to the previous or next note. Tap the Mail icon, and a new email message opens in Mail, with the note's text appearing in the message body. Tap the Trash icon, and you'll be offered the option to Delete Note or Cancel.

note **Regrettably, email is currently the only way to get notes out of your iPhone; they don't transfer from the iPhone to your computer when you sync the iPhone.**

To view a list of all your notes (**Figure 8.13**), tap the Notes icon in the top-left corner of the screen. Each note will be titled with up to the first 30 characters of the note. (If you've entered a return character after the first line, only the text in that first line will appear as the note's title.) Next to each note is the date of its creation (or time, if created that day). Time and date information also appears at the top of each note.

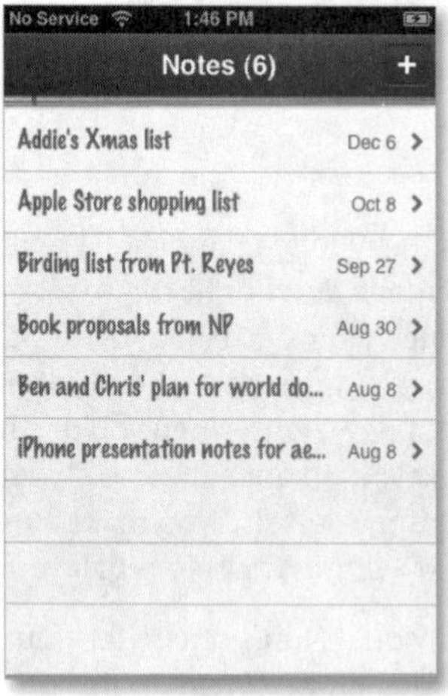

Figure 8.13 Notes application.

9

Tips and Troubleshooting

Compared with just about any other mobile phone you've owned, the iPhone is a dream of intuitive design and ease of use. Yet nothing in this world (save you, dear reader, and I) is completely foolproof or infallible, which is why this chapter is necessary.

Within these pages, I offer ways to get things done more expeditiously, provide hints for operating the iPhone in ways that may not seem obvious, and (of course) tell you what to do when your iPhone does the Bad Thing and stops behaving as it should.

Getting Tipsy

I've sprinkled tips and hints throughout the book, but I saved a few good ones for this chapter. In the following sections, I show you how to control text better, manage your iPhone's battery, sync your iPhone more efficiently, and pick up a valuable resource for taking your iPhone way beyond Apple's envisioned limits.

The word on text

If one iPhone feature frustrates the greatest number of people from the get-go, it's text entry. These tips will help make you a better iPhone typist.

Keep going

I've mentioned this before, but I'll say it again here: Typing on the iPhone's keyboard isn't like typing on your computer keyboard, a process in which you type, make a mistake, backspace to correct the mistake, and continue typing. Use that technique on the iPhone, and you'll go nuts making the constant corrections.

Typing the first letter correctly is important, as mistyping that first letter is likely to send the iPhone's predictive powers in the wrong direction. But after that, get as close as you can to the correct letters and continue typing even if you've made a mistake. More often than not, the iPhone's predictive typing will correct the mistake for you (**Figure 9.1**).

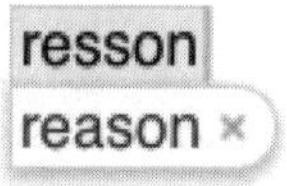

Figure 9.1
More often than not, the iPhone knows what you meant to type.

Sure, you may need to go back and correct a word or two in a couple of sentences by pressing and holding the display to bring up the magnifying-glass icon, but doing this for two mistakes is far more efficient than retyping half a dozen words.

Adjust your aim

When you start typing on your iPhone, you'll discover that your aim is off. Because I'm right-handed, for example, I tend to tap the right edge of a letter and often type the letter to its right instead. The iPhone, however, likes its letters tapped right in the middle. Similarly, I aim a little low in lists.

If you find that you're missing more often than you hit, consciously type to the opposite side of the key or command, or try typing with the entire pad of your finger rather than just the tip. Chances are that you'll nail what you're trying to type.

Move to the correct letter

In one specific instance, you'll need to type as carefully as possible: when you're entering a password. As I've mentioned elsewhere, for security reasons the iPhone enters dots in a password field rather than characters, so you don't have the luxury of going back to correct your work, because you can't see where you've made a mistake.

For this reason, when you're entering passwords (or just typing carefully), tap a character and wait for the letter to pop up on the display. If you've hit the wrong character, keep your finger on the display and move it to the correct character. Only when you release your finger does the iPhone accept the character.

Adjust the dictionary

Irked because the iPhone invariably suggests *candle* when you intended to type *dandle* (**Figure 9.2**)? You have the power to modify the iPhone's built-in dictionary. If you type *d-a-n-d-l-e*, but the iPhone displays *candle*, simply tap the *candle* suggestion, and it disappears. Then finish typing.

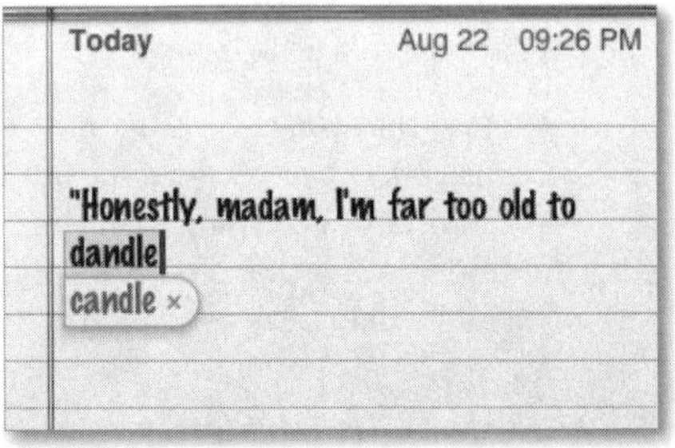

Figure 9.2 Correct the dictionary by tapping incorrect suggestions.

When you next get a good way into typing *dandle*, the iPhone will propose it as the word to use. When it does, just tap the spacebar to autocomplete the word. The iPhone's not stupid, so it won't suggest *dandle* when you next type *candle*, but it may not autocomplete *candle* that first time. In subsequent entries, however, it probably will.

Avoid unnecessary capitalizations and contractions

The iPhone tries to make as much sense as possible from your typing. When it's willing to, let it carry the load. For example, you probably won't type the letter *i* all by itself unless you mean *I*. The iPhone knows this and will make a lone *i* a capital *I*. Similarly, type *ill*, and even if you're trying to say that you're not feeling well, the iPhone will suggest *I'll*. Conversely, if you're feeling fine, the iPhone allows you to type *well* without suggesting *we'll*. Knowing that both *its* and *it's* are common, the iPhone will never suggest the contraction.

Rule of thumb: When a word that can also be spelled as a contraction is tossed at the iPhone, it will suggest the more commonly used word (**Figure 9.3**).

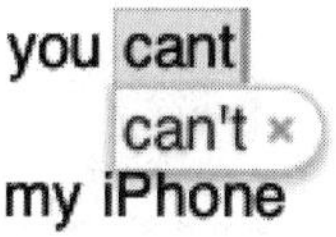

Figure 9.3 You can often skip the apostrophes when typing on the iPhone.

Use Pogue's punctuation tip

The New York Times' technology columnist, David Pogue, revealed this tip scant days after the iPhone was released, and in doing so, he proclaimed that other technology writers would use it in a heartbeat because it's so good.

Darn tootin', say I. And it goes like this:

You may find it distracting to have to tap in and tap out of the iPhone's number/punctuation screen whenever you want to add a stray comma or type *9* rather than *nine*. This dance isn't necessary. Just tap and hold the .?123 icon in the bottom-left corner of the keyboard. While holding down your finger, drag to the punctuation symbol or number you want to type. When that item is selected, let go. The keyboard will return to the alphabetical keyboard.

Regrettably, this trick doesn't work in reverse: You can't tap over to the number/punctuation screen, tap and hold the ABC icon, choose a character, and expect to remain with the number/punctuation keyboard. Try as you might, you'll switch to the alphabetic keyboard.

Power management

Wonderful as it is to have a mobile phone that can play full-length movies, you do *not* want to board a cross-country flight, enjoy the latest Harry Potter flick on your phone, jump off the plane with the expectation of alerting a key client to your arrival, and be greeted with a dead battery. Power can be paramount in such situations. To help ensure that your battery will still have something to offer, try these tips.

Treat it right

Your iPhone's battery performs its best in these conditions:

- **It's warm.** Lithium batteries perform best when they're run at around room temperature. If they get cold—below 24° F—they don't hold a charge as long.

- **But not too warm.** Running a cool iPhone won't damage the battery, but storing it somewhere that's really hot—say, your car's glove compartment when it's 95° F outside—can. Also, the iPhone gets warm when you charge it and extra-warm when you charge it in a case. Therefore, don't leave your iPhone in a hot place, and remove it from a case before charging.
- **It's used at least once a month.** I can't imagine owning a mobile phone that you never unplug from a power source, but it takes all kinds to make a world. The iPhone's battery likes to have its little electrons banged around at least once a month. Do so by unplugging it and making a few calls or playing a few tunes.

Lock it

The iPhone isn't supposed to do anything unless you touch its display or push its Home button, but you might accidentally do one thing or the other if the iPhone is rattling around loose in your pocket or pocketbook. Rather than projecting all 216 minutes of *Lawrence of Arabia* to the inside of your pants pocket, quickly press the On/Off button to lock your iPhone.

Turn off Sound Check and EQ

The iPod features Sound Check and EQ (equalizer) require more processing power from your iPhone, in turn pulling more power from your battery. If you've applied EQ settings in iTunes to the tracks that will play on your iPhone, you must set the iPhone's EQ setting to Flat, which essentially tells the iPhone to ignore any EQ settings imposed by iTunes. To make EQ Flat, choose Settings > iPod > EQ, and tap Flat in the EQ screen.

Turn on airplane mode

If you don't need to make or receive calls or to use the iPhone's Internet capabilities, switch the iPhone to airplane mode (go to the Settings screen, and toggle the Airplane Mode switch to Off).

note **You'd be surprised how much power the iPhone sucks when it's just sitting around waiting to receive a call. In one music-play-time test, I switched on airplane mode, and the iPhone played music for more than 37 hours straight.**

Turn off Wi-Fi

Although it doesn't save you as much juice as switching the iPhone to airplane mode, turning off Wi-Fi can help you get more life from your iPhone charge. To turn off Wi-Fi, go to the Settings screen, tap Wi-Fi, and flip the toggle switch to Off.

Plug it in

If you're accustomed to the way an iPod works, you may be under the impression that when you jack your iPhone into your computer's USB port, you can't use it. Wrong. When it's plugged into its power supply, your computer, or an accessory device that supplies power, the iPhone is completely usable. Make calls, watch movies, surf the Net, get email—everything works.

note **If you plug the iPhone into an audio device, such as an FM transmitter or speaker system, you'll be prompted to switch on airplane mode, as using the iPhone in this fashion with the telephone features switched on can cause ugly audio artifacts to come out of your speakers.**

Use the Dock

Those who are accustomed to the iPod often charge the iPhone by plugging it into a computer. This method isn't the best charging solution, though, because if your computer goes to sleep, the iPhone won't charge; it charges over USB only when the computer it's attached to is up, running, and awake.

Apple provides a small power supply with the iPhone. If you get into the habit of using the power supply, you'll never be caught unawares by an iPhone that's failed to charge.

Take it outside

The iPhone gets its charge via the same Dock connector as the iPod, so you can use the same power accessories that work with an iPod to charge your iPhone. That car charger you purchased to juice your iPod on a road trip, for example, works with the iPhone too. Ditto for the chargers that fit those special airplane receptacles. Ditto once again for the external battery packs sold by outfits such as Battery Geek (www.batterygeek.net). Granted, these battery packs can be bulky, but the best of them—the small bricklike rechargeable lithium-ion (Li-ion) batteries—can add dozens of hours of media play time to your iPhone.

Don't fret

Like all mortal things, the iPhone's battery will die eventually. You have more important things to do in life than micromanage your iPhone's battery. Enjoy your iPhone, and charge it when it's most convenient for you.

iPhone Battery: How Long and How Much?

When the iPhone was unveiled, many people were worried because the iPhone doesn't have a removable battery. How long will the battery last before it gives up the ghost? And will you have to buy a new phone when the battery dies?

To answer the first question, Apple claims that after 400 full charge cycles—that's a charge from dead to fully charged—the iPhone's battery will function at approximately 80 percent of its original capacity.

As for the second question, just as you can with an iPod (another device that's not designed for easy battery replacement), you can have your iPhone's battery replaced. If the iPhone is out of warranty—meaning that it's over 1 year old and you haven't added an AppleCare Protection Plan, which extends the hardware repair coverage to 2 years—Apple will do the job for $79 plus $6.95 shipping. The job takes 3 days. During that time, Apple will give you a loaner phone for $29 if you take your iPhone to an Apple Store. (The loaner phone is not available with mail-in service.)

By the time you read this book, you should see several third-party vendors jumping into the iPhone battery-replacement business, just as they did for the iPod. It's worth noting, however, that although some vendors may offer a "user-replaceable" battery, successfully replacing an iPhone's battery on your own is more than a little challenging. The battery is soldered in place. Unless you're *very* good with a soldering iron or couldn't care less about destroying your iPhone, have a professional do the job.

Sync different

Ask Apple about syncing your iPhone, and the answer you get is simple: One iPhone, one computer. But that answer's not entirely correct. To avoid that sinking feeling, keep these syncing tips in mind.

Sync to multiple computers

Although the iPhone lacks the iPod's manual-syncing option, you can sync your iPhone with different computers. The trick is that in nearly all cases, each computer will sync a different kind of media. You can sync music and videos from Computer A, photos from Computer B, podcasts from Computer C, and contacts and calendars from all three.

For this technique to work, you must enable the sync option in iTunes for just the media you intend to sync from a particular computer. So on Computer A, enable just the Sync Music, Sync TV Shows, and Sync Movies options. On Computer B, uncheck these options but check Sync Photos. Disable all these sync options on Computer C but enable the sync option for podcasts.

You can add contact, calendar, mail-account, and bookmark data from all these computers to a single iPhone. To do so, follow these steps:

1. Click the Info tab in iTunes' iPhone Preferences window, and enable the sync options you want—Sync Address Book Contacts and Sync iCal Calendars, for example.

2. In the Advanced area at the bottom of this window, where you see *Replace Information on This iPhone*, do *not* enable the options for contacts and calendars.

3. Click Apply.

   A dialog box will appear, asking whether you'd like to replace the information on your iPhone with the information on the currently connected computer or to merge the data on this computer with the data currently on the iPhone.

4. Click Merge.

   The chosen information on the computer will be merged with the existing information on the iPhone.

The agnostic iPhone

If you plug a Mac-formatted iPod into a Windows PC, you'll be told that you must restore the iPod—meaning that you must erase all the data on it and format it for Windows. Not so with the iPhone. You can plug it into a Mac and then into a Windows PC (or vice versa), and iTunes won't gripe—other than to offer to replace its media with the contents of the current computer's iTunes Library, as will happen when you plug your iPhone into any computer it wasn't last synced with.

Sync from your old phone

You probably have some treasured contacts and events on your old mobile phone. How do you get them from the old phone to your new iPhone? In the program you use to manage your old phone's data, export those contacts and events in a file format that's compatible with the iPhone (and with today's address-book and calendar applications), and import them into your iPhone-compatible PIM applications.

Specifically, export your contacts as vCards and your events as .ics files. All modern PIM applications support these formats, including Palm Desktop; Apple's Address Book and iCal; Microsoft's Entourage, Outlook, and Outlook Express; and Windows' Address Book and Calendar.

After you've exported the contacts and events, bring them into the applications that the iPhone uses for contacts and calendar data.

The Frankenphone

If a picture is worth a thousand words, this picture (**Figure 9.4**) of my iPhone's Home screen just saved you from poring over a couple more pages. As you can see, my phone's Home screen looks anything but normal, as it's supposed to contain 16 icons rather than 20.

Figure 9.4 You're right—your iPhone's Home screen *doesn't* have these icons.

The presence of Installer, Launcher, Lights Off, and Frotz (and even more applications stored behind the scenes) is due to my phone's having been hacked. All-around smart guy and good Joe (and fellow Peachpit author) Ben Long scoured the Web to learn how to hack my iPhone. You see these clear screen shots taken directly from the phone rather than fuzzy photos because of Ben (and Erica Sadun, who created the Screenshot iPhone application).

note **Rest assured that as I type these words, hackers and other third parties are working on additional (and easier) ways to install cool stuff on your iPhone. Ambrosia Software (www.ambrosiasw.com), for example, showed me its $15 iToner utility for installing customized ringtones on the iPhone without requiring you to hack the phone.**

tip **Apple may bar the door to such hacks in the future or, better yet, provide developers a way to install their applications on the phone. Until that day arrives, allow me to direct your attention to an iPhone hacking tutorial Ben has assembled (and hopes to keep updated). Travel to www.geeklagoon.com for a link to Ben's latest iPhone hacking work.**

Troubleshooting

The iPhone may be an engineering marvel, but even engineering marvels get moody from time to time. And when your iPhone misbehaves, you're bound to be in a hurry to put things right. Allow me to lend a hand by suggesting the following troubleshooting techniques.

The Four Rs

In the following pages, I repeatedly refer to four troubleshooting techniques: resign, restart, reset, and restore. In order of seriousness (and desirability), they are:

- **Resign.** Force-quit the current application by holding down the Home button for about 6 seconds. This step should get you out of a frozen application and return you to the iPhone's Home screen.
- **Restart**. Turn the phone off and then on. Hold down the Sleep/Wake button until a red slider appears that reads *Slide to Power Off*. Slide the slider, and the iPhone shuts off. Now press the Sleep/Wake button to turn on the iPhone.
- **Reset.** Press and hold the Home and Sleep/Wake buttons for about 10 seconds—until the Apple logo appears—and then let go. This step is akin to resetting your computer by holding down its power switch until it's forced to reboot.
- **Restore.** Plug your iPhone into your computer, launch iTunes, select the iPhone in iTunes' Source list, click the Summary tab, and click the Restore button. This step wipes out all the data on your iPhone and installs a clean version of its operating system.

Fortunately, iTunes makes a backup of your information data (contacts, calendar events, notes, and so on) when it syncs the iPhone. After restoring the iPhone, you'll be asked whether you want to restore it from this previously saved data. You do.

The basics

If your iPhone acts up in a general way—won't turn on, won't appear in iTunes, or quits and locks up—try these techniques.

No iPhone startup

Is your phone just sitting there, with its cold black screen mocking you? Try charging it with the included charger rather than a USB 2.0 port. If you get no response after about 10 minutes, try another electrical outlet. Still nothing? Try a different iPhone cable.

Still no go, even though you've had that iPhone for a long time and use it constantly? The battery may be dead (but this shouldn't happen in your first year of ownership, regardless of how much you use the phone).

No iPhone in iTunes

If your iPhone doesn't appear in iTunes when you connect it to your computer, try these steps:

1. Make sure your iPhone is charged.

   If the battery is completely dead, it may need about 10 minutes of charging before it can be roused enough to make an iTunes appearance.

2. Be sure the iPhone is plugged into a USB 2.0 port.

   Your computer won't recognize the phone when it's attached to a USB 1.0 port.

3. Plug your iPhone into a different USB 2.0 port.

4. Unplug the iPhone, turn it off and then on, and plug it back in.

5. Use a different iPhone cable (if you have one).
6. Restart your computer, and try again.
7. Reinstall iTunes.

Unresponsive (and uncooperative) applications

Just like the programs running on your computer, your iPhone's applications can act up, freezing or quitting unexpectedly. You can try a few things to nudge your iPhone into action. If the first step doesn't work, march to the next.

1. Resign from the application.

   If an application refused to do anything, it's likely frozen. The only way to thaw it is to force it to quit. Press and hold the Home button until you return to the Home screen.

2. Clear Safari's cache.

   If you find that Safari quits suddenly, something in its cache may be corrupted, and clearing the cache may solve the problem. To do so, tap Settings in the Home screen; then tap Safari; and in the Safari Settings screen, tap Clear Cache.

3. Reset the iPhone by holding down the Home and Sleep/Wake buttons until you see the Apple logo.
4. On the iPhone, go to the General setting; tap Reset; and then tap Reset All Settings.

   This step resets the iPhone's preferences but doesn't delete any of your data or media.

5. In that same Reset screen, tap Erase All Content and Settings (**Figure 9.5**).

note

This step vaporizes not only the iPhone's preferences, but its media content as well. Before doing this, try to sync your iPhone so that you can save any events, contacts, bookmarks, and photos you've created.

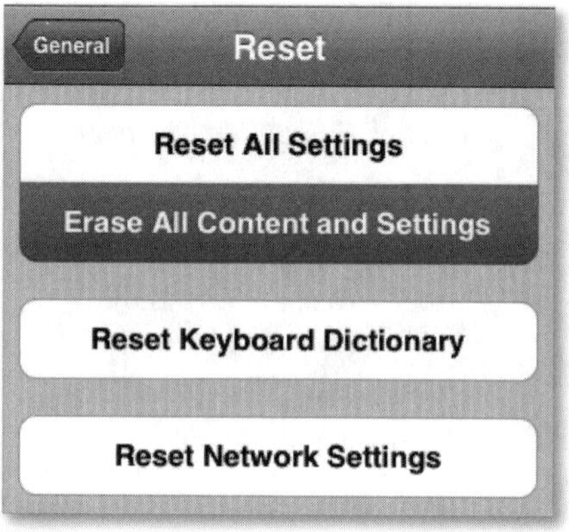

Figure 9.5 Erasing all the content and settings from your phone is the next-to-last resort.

6. Restore the iPhone.

Sunk by sync

Data should move smoothly between your computer and your iPhone, but it doesn't always. Try these fixes.

iPhone runs out of free space

You may see an error message indicating that the iPhone doesn't have enough free space to sync all the data and media you've selected in iTunes. If you inadvertently unplugged the iPhone during a sync, some data may be left on it, taking up space. To remove this excess data, disable music and photo syncing in iTunes, and click the Apply button to sync the phone. This step should erase the data. Now enable music and photo syncing, and click Apply again to sync your media.

Another possibility is that you're simply asking the iPhone to suck up too much media. You may have accumulated a lot of video podcasts since your last sync, for example, and the phone just doesn't have room for all of them. Try disabling files that take up a lot of storage—TV shows, movies, and video podcasts—and then sync the phone. Look at the Capacity bar to see how much space remains, and choose media based on that remaining space.

Photos won't sync

If your iPhone is locked with a passcode, photos you've taken with its camera won't sync to your computer. Unlock your iPhone, and the pictures will transfer.

Yahoo Address Book won't sync

If you receive an exception error when attempting to sync Yahoo Address Book contacts, one of the contacts in your regular address book may have an ill-formatted address. If you have *billybob.example.com* instead of *billybob@example.com*, you may see this error. Check the addresses in the program that holds the contacts you sync with your iPhone.

Mail issues

Are your attachments not opening? Is the iPhone refusing to send your mail? Are you getting far too many offers for questionable nostrums and shady real estate deals? Read on for solutions.

Can't read attachments

You can read certain kinds of documents that arrive as attachments in email messages—specifically, Microsoft Word, Microsoft Excel, PDF, JPEG, and text files. But Word, Excel, and PDF files won't open unless they carry the proper extensions: .doc, .xls, and .pdf, respectively. Also, if the message body is formatted as rich text (RTF) and includes an attachment, you won't be able to read the attachment. Try forwarding the message to yourself. This method should convert the rich text to plain text and allow you to view the attachment.

Can't send mail

In Chapter 4, I mention that if your ISP uses an SMTP port other than 25, you should enter a colon and then append the port number to the end of the SMTP address, as in *smtp.examplemail.com:587*.

If this situation isn't the issue, try configuring the email account on your iPhone rather than syncing it from iTunes. To do so, copy the settings you'll need from your email client (account name, password, POP, SMTP or IMAP information); delete the account that was synced from iTunes; go to the Settings screen; turn off Wi-Fi; and then tap the Mail setting.

In Mail, tap Add Account; configure the account; and tap Save. The iPhone will attempt to confirm your account over EDGE. If it does so, you're good to go. Turn Wi-Fi back on.

If you can't send mail because your ISP prohibits you from *relaying* (sending mail through another ISP, as you might when connected to a Wi-Fi network other than your own), you can use AT&T's cwmx.com outgoing server to send mail over EDGE. Or consider adding a free Gmail (http://mail.google.com), Yahoo (http://mail.yahoo.com), or AOL (http://mail.aol.com) account and sending mail via its server.

Can't cope with spam overload

The iPhone's Mail program offers no spam filtering. If your computer's email client removes the bulk of the spam you receive, you'll be shocked when you download your first batch of mail on the iPhone, because it's likely to be choked with spam.

If your ISP can't impose some kind of filtering on your email so that the spam doesn't reach you in the first place, sign up for a free Gmail account, and switch to it for email you intend to receive on your iPhone. Gmail has great spam filtering, so you'll get just the mail you want without the excess junk. (You can also configure Gmail to forward mail from other accounts through your Gmail account and remove the spam in the process.)

Index

J

K

L

R

S